A FIELD GUIDE FOR SCIENCE WRITERS

A Field Guide for Science Writers

SECOND EDITION

EDITED BY

Deborah Blum

Mary Knudson

Robin Marantz Henig

2006

OXFORD
UNIVERSITY PRESS

Oxford University Press, Inc., publishes works that further
Oxford University's objective of excellence
in research, scholarship, and education.

Oxford New York
Auckland Cape Town Dar es Salaam Hong Kong Karachi
Kuala Lumpur Madrid Melbourne Mexico City Nairobi
New Delhi Shanghai Taipei Toronto

With offices in
Argentina Austria Brazil Chile Czech Republic France Greece
Guatemala Hungary Italy Japan Poland Portugal Singapore
South Korea Switzerland Thailand Turkey Ukraine Vietnam

Published by Oxford University Press, Inc.
198 Madison Avenue, New York, New York 10016

www.oup.com

Library of Congress Cataloging-in-Publication Data
A Field guide for science writers : the official guide of the National Association of Science Writers / edited by Deborah Blum, Mary Knudson, Robin Marantz Henig.—2nd ed.
p. cm.
Includes index.
ISBN-13 978-0-19-517498-4; 978-0-19-517499-1 (pbk.)
ISBN 0-19-517498-4; 0-19-517499-2 (pbk.)
1. Technical writing—Handbooks, manual, etc. I. Blum, Deborah. II. Knudson, Mary. III. Henig, Robin Marantz.
T11.F52 2005
070.4'495—dc22

2005001267

9 8 7 6 5 4 3 2
Printed in the United States of America
on acid-free paper

FOREWORD

TIMOTHY FERRIS

Science, though young, has already transformed our world, saving over a billion people from starvation and fatal disease, striking shackles of ignorance and superstition from millions more, and fueling a democratic revolution that has brought political liberty to a third of humankind. And that's only the beginning. The scientific approach to understanding nature and our place in it—a deceptively simple process of systematically testing one's ideas against the verdict of experiment—has opened limitless prospects for inquiry. There is no known limit to the knowledge and power that may, for better or worse, come within our grasp.

Yet few understand science, and many fear its awesome power. To the uncomprehending, the pronouncements of scientists can sound as opaque as the muttered spells of magicians, and the workings of scientific technology resemble, as the French say of the law, a machine that cannot move without crushing someone. Technophobes warn that science must be stopped before it goes "too far." Religious fundamentalists enjoin the righteous to study only one (holy) book, consulting what Galileo called the book of nature only insofar as it serves to confirm their beliefs. Fashionable academics teach that science is but a collection of socially conditioned opinions, as changeable as haute couture. Popular culture is so suspicious of science that, according to one study, scientists portrayed in American feature films are more likely to be killed by the last act than are members of any other profession, including Western gunslingers and Mafia hit men.

The cure for fear and loathing of science is neither propaganda nor persuasion but knowledge—conveyed, preferably, in stories that capture and reward an audience's attention. Science writers, whose work involves crafting such stories, are few in number, relatively unheralded, and often underestimated: Like sportswriters and business journalists, they are too often assumed to be mere interpreters or translators rather than "real" writers, as if crafting an accurate, evocative paragraph about biochemistry or quantum physics were less of an achievement than doing the same thing when the subject was a lotus blossom or a love affair. But we science writers also enjoy certain advantages. We have what are, in many respects, the best stories to tell—the most momentous, important, and startlingly original stories, as you will find demonstrated by the writers collected here. Plus, science writers tend to be generous in spirit. "Interested in writing about science?" reads the subtext of this rich and readable book. "Come on over and give it a try!" Heed their counsel, accept their invitation, give it your best shot, and I'm betting you'll never go back.

EDITORS' NOTE

In the eight years since publication of the first edition of *A Field Guide for Science Writing*, much about the world has changed. Science writing has changed, too. Once the province of nerds and the nerds they wrote about, the field has evolved, becoming at once more esoteric—because science itself has become more esoteric—and more a part of daily life. Some of the leading issues in today's political marketplace—embryonic stem cell research, global warming, health care reform, space exploration, genetic privacy, germ warfare—are informed by scientific ideas.

Never has it been more crucial for the lay public to be scientifically literate. That's where science writers come in. And that's why it's time for an update to the *Field Guide*, already a staple of science writing graduate programs across the country.

When we first undertook this venture in the mid-1990s, it was something new for the National Association of Science Writers. From its beginnings in 1934 as an old boys' club of about a dozen science writers, NASW is today a professional organization with nearly 2,500 members. As the organization has grown, so has the profession—and now more than ever we need to be clear about what the profession is all about.

Writing well about science requires, first of all, bridging the jargon gulf, acting as translators between the sciencespeak of the researchers and the short attention spans of the public at large. But great science writing doesn't stop there. You can paint an awesome picture of space exploration with all its glittering astrotoys, but you also have a responsibility to probe its failures. You can point out the benefits of genetically modified crops or the mapping of the human genome, but you also must explore their potential to do harm. It's not enough to focus on the science itself; the best reporting also discusses safeguarding the public from the risks of the new knowledge and talks about the cost of Big Science and who has to pay for it.

The academic community has recently recognized how important it is for science writers to become more sophisticated, knowledgeable, and skeptical about what they write. More than 50 institutions now offer training in science writing. In addition, mid-career fellowships for science writers are growing, giving journalists the chance to return to major universities for specialized training. We applaud these developments, and hope to be part of them with this new edition of the *Field Guide.*

In these pages, we've assembled contributions from a collection of experienced science writers who are every bit as stellar as the group that contributed to the first edition of the *Field Guide.* When we editors thought about who would be best to contribute a particular chapter on writing for a particular medium—newspapers, magazines, trade journals, books, broadcasting, or the Web—or in a particular style—explanatory, investigative, narrative, essay, or what one contributor calls "gee whiz science writing"—we began by asking the top practitioners of that medium or that style. And guess what—they said yes! So what we have here are essays written by the very best in our profession. Their voices differ from one chapter to another, but that's what we wanted—a book that feels conversational and wise, a chance to pull up a chair and sit beside a kindly aunt or uncle who can tell you how it's done.

These wonderful writers have written not only about style, but about content, too. There's so much information to impart—some say there are more scientific articles published in the United States in a single year than were published from Gutenberg's day through World War II—that negotiating the morass can be especially daunting for a newcomer. So we asked the leaders of our profession to describe how they work their way through the information glut to find the gems worth writing about. As you can see from the table of contents, we've asked them to describe how they cover subjects ranging from astronomy to zoology, from the smallest microbe to the universe itself. We also have chapters that provide the tools every good science writer needs: how to use statistics, how to weigh the merits of conflicting studies in the scientific literature, how to report about risk. And, while we're at it, how to write.

As we put all these pieces together, we noticed two themes that kept recurring. Both of them seem to capture science writing at the beginning of the twenty-first century. The first relates to narrative. Over and over again, our authors advise you to look for the story, the narrative arc, that will compel your readers to stick around to find out about the science. This has always been a good idea—think back to one of the earliest examples of best-selling science writing, *Microbe Hunters* by Paul de Kruif, which has been continually in print since 1926. What's new, however, is that more and more writers are seeing the brilliance of doing what de Kruif did, presenting science as one great big storybook adventure.

The second recurring bit of advice concerns balance. Traditional journalism aims for objectivity by including representatives of both sides of any debate. But in many of the most vigorous debates in science, looking for contrary views would do your readers a disservice. There's no need to quote from the fringe groups—people who insist that HIV doesn't cause AIDS, or who don't believe in evolution, or who think Earth is flat—just because they exist. More than in any other field of reporting, balance in science writing requires something other than just providing an equal number of column inches to quotes from each side. Balance in science writing requires authorial guidance; it requires context, and knowing when certain points of view simply need to be ignored.

The science writers who contributed the bookends for the *Field Guide*, the foreword and the epilogue, are among the most luminous practitioners of the craft. Each of them graciously set aside his other obligations to take the time to think about our profession's particular strengths and challenges, hoping to illuminate the recondite corners of science writing in a way that will help the next generation. We would like to offer here a thank you to Tim Ferris and Jim Gleick, two men who have spent their careers elevating science writing by glorious example. Tim is the author of such brilliant books as *Coming of Age in the Milky Way* (1988) and *The Whole Shebang* (1997) and was once described as writing "as if brushed with stardust." Jim, whom one critic called the "consummate craftsman," writes books that are equally brilliant, including the bestsellers *Chaos* (1987) and *Genius* (1992), as well as *Isaac Newton* (2003), a finalist for the Pulitzer Prize.

While we're expressing appreciation, we'd like to thank all our other contributors, too, whose compensation was so small as to make their work for us essentially voluntary. They were entirely professional at every point of the process, responding with grace and speed to editorial direction that could have been awkward, coming as it did not only from colleagues and friends, but from a trio of us. Thanks, guys—you made it easy.

Thanks, too, to Joan Bossert, our editor at Oxford University Press, for seeing the need to update the *Field Guide* and for enthusiastically getting behind the project, as well as to her assistant Jessica Sonnenschein. Thanks to Mary Makarushka, whose sharp organizational skills kept the three of us on track during this book's assembly, and to Diane McGurgan of NASW, who always put in the extra effort on our behalf. And thanks to the organizations that provided much-needed financial support to see this project through: the Alfred P. Sloan

Foundation, the Council for the Advancement of Science Writing, and the National Association of Science Writers.

We were privileged to design this book and guide it to completion, a project made better by a warm camaraderie. We hope this *Field Guide* will help a new generation of science writers embrace our profession with enthusiasm, tenacity, and sophistication. And we hope you have a lifetime of fun doing it.

DEBORAH BLUM
MARY KNUDSON
ROBIN MARANTZ HENIG

CONTENTS

Part One

Learning the Craft

■ ■ ■ ■ ■

To you students who are aspiring science writers and to science and medical writers just starting out, welcome to science writing boot camp. How I wish I could have attended one! My life changed the day my editor unexpectedly told me that I was the new medical writer at the *Baltimore Sun*. The previous long-time medical writer had left on very short notice, and I was stuffed into a beat that had to be filled; overnight I went from being a generalist to being a specialist in a city that was home to the world-famous Johns Hopkins Medical Institutions, had a large and growing University of Maryland Hospital and School of Medicine, and was a short drive to the National Institutes of Health.

Never having covered medicine or science, I remember desperately trying to learn some of the scientific vocabulary on my way to my first science writers meeting, put on by the American Cancer Society. Once there, I was properly intimidated by the depth of knowledge reporters commanded as they grilled scientists who had made presentations. The best reporters seemed, from the framing of their remarks, to know as much about the subject as the scientists they were questioning. By comparison, I felt so not ready even to be at a cancer conference asking questions and deciding what may be a daily story. I experienced what a staggering challenge it is to get thrown in and have to start from scratch being a medical or science writer.

You may be about to jump in, too. Go ahead. I promise it gets easier as you develop news judgment, background knowledge, and very good sources. You get to know the territory. You come to know from extensive reading, reporting, and networking with well-connected smart sources what is big news and what is worth watching. And before you know it, you're one of those journalists standing up asking pointed, incisive questions. You're going to have a lot of fun!

And so part I of the *Field Guide* is especially for you. The authors, all masterful writers, will drill you in the basics of getting started as a science writer, from finding story ideas and sources to reporting accurately and writing well.

Then at the end of this part, two eloquent writers will take you to the next level, sharing lessons they have learned about how to pursue and write a story that is a standout, notable for its depth of reporting, style, and voice.

Begin by reading extensively, Phil Yam advises in chapter 1. Read science stories in the media and scientific papers written by scientists in journals. If you are a student, you should be able to access PubMed, Lexis Nexis, and other databases through your university. Find out from a librarian how to connect your home computer into the university system. You can access PubMed and many other databases direct through your own Internet connection, but you're more likely to get full-text articles from more journals by routing through your university, which subscribes to the journals. It is crucial for you to build sources, and Phil gives tips for doing that.

Two of the most challenging responsibilities you take on as a science writer are reading journal articles and really understanding statistics. You need to know how to read a scientific paper published in a journal to see if it is worth writing about. To help in making your decision, it's important to understand statistics and know what questions to ask scientists about how their studies were conducted and what the results mean.

In chapter 2, Tom Siegfried explains the importance of peer-reviewed journals and names the most widely read ones. He walks you through how to read a scientific paper critically and assess its worth, and gives commentary on the embargo system about which all science writers must be aware.

Do statistics scare you, leave you feeling ignorant, ashamed, disoriented? You're in the right place. In chapter 3, Lew Cope tells you what questions to ask "to separate the probable truth from the probable trash." He also explains five principles of scientific analysis and defines those oft-used terms *statistical probability* and *statistically significant.* With the information in this chapter, you'll be able to go well beyond asking scientists to explain their findings in English. You'll be equipped to ask challenging questions that test whether the scientific skeleton on which the study was built supports its conclusions.

In chapter 4, some of us who teach science writing at universities share techniques for writing well about science. This is sort of a smorgasbord of tips you can use immediately. "Use transitions. A story has to flow. Leaping from place to place like a waterstrider on a pond will not make your prose easy to follow," Deborah Blum charmingly advises. And in doing so, she sneaks in a great little simile.

With the basics behind us, in chapter 5 Nancy Shute discusses "Taking Your Story to the Next Level." This is a very thought-intensive effort, and once you get an idea for a big story, you begin with extreme measures of reporting. "I like to think of journalism as bricklaying," Nancy writes, "a noble craft, but a craft all the same." She gives four hallmarks of a great story: "a good story idea, meticulous reporting, great characters, and the right perspective." When they are all put together, she writes, "the results can be riveting."

Nancy uses a story by Atul Gawande to depict this riveting result. Gawande, a practicing surgeon who is also one of us, a journalist, narrative writer, and essayist, could just as well be held up as an exemplar for the topic that closes out part I: writing with a voice and style. One quality that resonates from all his stories is honesty.

Style and voice are those qualities, elusive to define and teach, that, I think, makes a story professional and publishable. Your "personality on the page," David Everett calls them. In chapter 6 he gives us a recipe: "Style and voice flow from straightforward elements such as rhythm, punctuation, verb tense, word choice, sentence construction, adjectives and adverbs." The list continues and includes "larger artistic mysteries." It all sounds daunting, but don't worry. By the time you have arrived at this juncture in journalism where you are chiseling your personality on the page, you will know how to use all these tools. And developing your style and your voice will be the most fun of all.

MARY KNUDSON

1

Finding Story Ideas and Sources

PHILIP M. YAM

Philip Yam cut his journalism teeth as a staff writer for the independent *Cornell Daily Sun,* the morning newspaper in Ithaca, New York, while studying physics at Cornell University. A few years after graduating in 1986, he joined *Scientific American* as a copy editor. A year later, he became an articles editor, writing news stories and profiles in addition to editing scientist-authored material and the "Amateur Scientist" column. Then, in September 1996, he became the news editor. Phil was a science writing fellow at the Marine Biological Laboratory in Woods Hole, Massachusetts, and the Knight Foundation boot camp at MIT. The subject of prions provoked his interest enough to write a popular-science book, *The Pathological Protein: Mad Cow, Chronic Wasting, and Other Deadly Prion Diseases* (2003).

As a freelance or a staff journalist, you will face at some point dread and insecurity as you wonder if the story ideas you're about to pitch to an editor are any good. We've all been there. There is no formula for coming up with that novel angle or fresh topic. But certain approaches and strategies can help you hone your nose for science news and root out interesting stories editors will want.

First, scope out publications, both print and Web. If you've contemplated science journalism, then you have probably read the science and technology sections of major newspapers and leafed through the popular-science magazines on the newsstands.

Familiarize yourself with the weeklies, such as *New Scientist* and *Science News,* as well as the news section of *Science.* Gain a greater depth by, for instance, reading review-type articles, such as those that appear in *Scientific American*, *Nature*'s News and Views section, or the News & Commentary section of *Science*.

Check out clearinghouses for press releases, such as Newswise, Eurekalert!, and PRNewswire. They send periodic e-mail alerts and maintain searchable websites. Some require that you have a published body of work before granting you access to certain privileged information (such as the contact numbers of researchers). Others may require that you obtain a letter from an editor. You can also subscribe to mailing lists of media relations offices at universities, medical centers, and other research institutions and sign up for various industry newsletters.

When surfing the Web for science information, don't forget major government websites, such as those of the National Aeronautics and Space Administration, the National Institutes of Health, the National Institutes of Standards and Technology, and the Department of Energy, which manages the national labs. Besides weapons work, the DOE labs—including Los Alamos, Brookhaven, Oak Ridge, and Lawrence Livermore—conduct research in both physical and biological sciences. Other worthwhile online resources include listservs and Web logs, but keep in mind that the ideas there are not vetted as they are in journals. Plus, you have to have the patience to get past the ranting and raving that can obscure good postings. For beginning science journalists, it may be best to follow blogs of well-respected researchers.

You can also try fishing for stories directly from journals. Be warned, though, it takes an experienced eye to mine the vast numbers of papers with impenetrable titles published every month. Would you have guessed that "Lysosomotropic Agents and Cysteine Protease Inhibitors Inhibit Scrapie-Associated Prion Protein Accumulation" refers to certain drugs that could treat mad cow disease? Don't worry; nobody else did, either—until a year later in 2001, after another team reported similar findings and had the benefit of a press release issued by its university.

Despite the potential pitfalls, journal scoping is a way to get to a story no one else is likely to pursue. For the physical sciences, a popular place to look is www.arXiv.org, an online preprint library. There is no current analogue for the biological sciences, but I have found the National Library of Medicine's database of published articles, PubMed (www.ncbi.nlm.nih.gov/entrez/query.fcgi), to be useful. PubMed is a major resource for finding medical journal abstracts and many full-text articles, and I feel more comfortable with an idea if it has generated legitimate papers in top-notch journals by recognizable authors.

If you are a university student, you should be able to access PubMed, Lexis Nexis, and other databases from your home computer by routing through your university library. Schedule a time to sit down with a librarian who can tell you how to link your home computer to these databases through Remote Access to University Libraries (RAUL) or some other system at no charge. The advantage to accessing medical journals through a university library is that the library subscribes to most journals you would want. So if you can't otherwise get more than an abstract, you can more likely get the full text of an article through the university library. Lexis Nexis is a quick way to find out whether a story you want to write has already been written in magazines, newspapers, or scientific newsletters, or to get background information on a subject that interests you.

Following the money can pay off as well, notes Christine Soares, a *Scientific American* editor and former writer and editor for *The Scientist.* As she puts it: "If a funding agency like the National Science Foundation creates a new program, or a national lab announces they've just tripled spending on some particular line of research, it could be a sign that the field has reached some critical mass and is worth looking into. This can mean slogging through the *Federal Register* and/or subscribing to assorted e-mail newsletters (for example, the American Society for Microbiology and the American Institute of Biological Science have 'funding alert' e-mails), but may occasionally pay off in a very early lead on a field that's going to be making news."

Prizes can also be an excellent source. The Nobels, announced in early October, are often the time when basic research takes the spotlight, although they are also often a time capsule of discoveries of a bygone decade. More up-to-date work is honored by the MacArthur Foundation, which focuses particularly on researchers who are young, working in a hot field, and not getting the grants afforded to more easily fundable topics. In part, that is how I came to ask contributing editor Marguerite Holloway to profile two investigators in 2004: Bonnie Bassler, a Princeton University biologist studying quorum-sensing in bacteria (how they decide to act depending on their numbers); and Deborah Jin, a physicist who created a new state of matter with ultra-cold atoms. The Albert Lasker Medical Awards often point the way to future Nobel Prize winners. Lesser known annual awards include the Kyoto Prize and the Lemelson—MIT prizes.

Keeping up with what's going on and learning which kinds of stories are most likely to make it in print, on the Web, or over the air will help you develop news judgment. Having such a background also helps in formulating novel angles and coming up with the day-after analysis that headline news often lacks. (As news editor, I encourage all writers to come up with deeper analyses.) The more you know what's going on, the better you will be at recognizing a good story when it comes along.

That's how I ended up being the first to write about the discovery of the Bose–Einstein condensate for *Scientific American* when I was an articles editor. The Bose–Einstein condensate (BEC for short) develops when a dense gas is trapped and chilled to a few billionths of a degree above absolute zero. Driven by the Heisenberg uncertainty principle—as the velocities of the gas atoms decrease, their positions become more unknown and must overlap—the atoms condense into one giant entity. Since 1925, when Indian physicist Satyendra Bose and Albert Einstein predicted it, physicists wondered if this quantum ice cube could indeed form. Creating the BEC was one of those long-sought goals of scientists that inspired a race among different groups.

In 1994 researchers managed to refine the refrigeration and trapping technology so that atoms could be chilled to where Bose—Einstein condensation is supposed to occur. Physicists began achieving ever lower temperatures—from thousandths to millionths to billionths of a degree above absolute zero. As I collected the various reports about the low-temperature records, I became convinced that someone would soon make the BEC. In May 1995, I got the go-ahead from my news editor to proceed with a story about the race, and I began in late May making phone calls to physicists at the Massachusetts Institute of Technology, the National Institute of Standards and Technology (NIST), and the University of Colorado at Boulder.

My second phone conversation with Eric Cornell of NIST took place on the afternoon of June 5, which turned out to be the day his team first made a BEC out of rubidium atoms. I remember thinking that I must be the only journalist in the world to know of the discovery and could actually break the story in a monthly magazine.

My excitement soon turned to frustration because Cornell and senior researcher Carl Weiman quickly decided that they wanted to publish their article in *Science.* The journal's embargo policy—shake fist now—scared the researchers out of continued talks with me. But I had enough information to write the story; my main worry was that they might retract their finding while our August issue went to press. Fortunately, except for a small detail I got wrong—the number of atoms trapped—things worked out: Our subscribers found out about the BEC in early July, a few days before the discovery made the cover of *Science* and the front page of the *New York Times.*

As is true for any kind of journalism, the best sources are people. If you studied science in college, you can tap old professors, teaching assistants, and even fellow students who have pursued science as a career. Just ask them what is the most interesting thing going on in their field right now.

Meetings are the most efficient way to connect with a lot of sources. The biggest, at least for the diversity of topics offered, is the annual meeting of the American Association for the Advancement of Science (AAAS), held in February. Typically, however, speakers at this meeting do not present a lot of new news, although the sessions can provide significant background information.

Smaller meetings are often a better bet; virtually every field, from anthropology to zoology, has associations or societies that hold meetings that are open to journalists. The American Physical Society (www.aps.org) holds its biggest meeting in March, when condensed-matter physicists gather to discuss the behavior of solids and liquids. About a month later comes the APS meeting covering most of the other branches, especially astrophysics and particle physics. Other subtopical meetings—for acoustics, nuclear, and optical, among others—are scattered around the country and the calendar. The American Chemical Society (www.chemistry.org) holds two national meetings a year, plus several regional meetings.

National meetings of societies are still large—the Society for Neuroscience (apu.sfn.org) November meeting draws around 25,000 researchers—and can easily overload your neural circuitry. The American Heart Association's annual meeting (scientificsessions.americanheart.org), also in November, is where the biggest news in cardiology is made.

To keep things manageable, set up an agenda before you actually get to a big meeting, preferably well before the airplane ride there. Look over the program and abstracts. Then map out which talks you want to attend. The invited talks are easier to grasp: Most of the contributed abstracts are by graduate students presenting their data to their immediate colleagues, and you have to be pretty familiar with the topics to appreciate them. Invited talks, however, can still be daunting. When covering the APS March meeting, I would call the speakers a couple of weeks beforehand and try to set up a meeting over coffee before or after their talks. That way, I had their undivided attention and could get all my questions answered, while also feeding my caffeine addiction. Away from the microphone, most presenters are more casual and accommodating. Don't overlook the organizers of panel talks themselves; they can provide impartial context.

Rather than hooking up with sources at official gatherings, you can request a private audience. Mariette DiChristina, *Scientific American*'s executive editor, recommends taking advantage of your location—especially if you happen to be where editors and other writers aren't. In her words: "A great way to find new news is to spend a day at a local research institution of your choice. You can start by contacting the public information officer and, ideally, you might set up a day or so of interviews. The PIO can help make recommendations about researchers whose work could be newsworthy, or you can make your own suggestions about people you'd like to see. Be clear about your intentions: You're a writer on the hunt for story ideas, which you hope to sell." You can't make any promises, but make it clear that you have every intention of placing a story in a media outlet.

Don't schedule more than one interview per hour, Mariette recommends, and "follow up later with the people you meet—to cultivate the relationship and to keep tabs on work that is progressing."

As in any good interview, pay attention to the details, which can sometimes lead to a better story. That's how *Scientific American*'s senior writer W. Wayt

Gibbs managed to break the story about the growth of new neurons in adult humans in 1998. Wayt had been following up on the research of Elizabeth Gould, a Princeton University biologist who made headlines in March 1998, with news that adult monkeys can grow new neurons. He contacted several researchers, many skeptical of the finding because of concerns about the experimental protocol. Among those whose input Wayt solicited was the Salk Institute's Fred "Rusty" Gage, who informed Wayt about his reservations while also saying that Gould wasn't necessarily wrong—a statement that makes an astute journalist's ears perk up.

In Wayt's words: "I sensed he was holding back and pressed him on the topic. He said that he had preliminary results that were very intriguing but couldn't talk about them yet and suggested I call him back in a few weeks." That tantalized us into killing the story about Gould's work—I had become the magazine's news editor by then—and finding out just what Gage was getting at. "I kept pestering Gage and at last in July he allowed me to come visit his lab at the Salk. We made an agreement that he would tell me all about his research, but I would not publish until he had submitted his paper for publication and gotten it through peer review." Wayt spent hours with Gage going over persuasive evidence more interesting than monkey brains, namely, that adult human brains can sprout new neurons, proving textbook dogma wrong.

To honor our agreement with Gage, we held off running the story for the next issue, and then again for the next. By September 1998, while I was lining up stories for the November news section, Wayt learned that the paper was finally in peer review at *Nature Medicine* and was being fast-tracked. So I decided to slot it as the top story for the November issue, which would appear in early October. As a courtesy, Wayt contacted *Nature Medicine* to inform the editor that we would be breaking the story.

We ended up catching some unfair flak for this—the NASW newsletter *ScienceWriters* chastised us in a story about uncontrolled embargo breaks. But *Nature Medicine* embargoed the story well after we had told them about our plans and had gone to press, so we didn't violate the journal's policy. Moreover, it would have been unfair to allow Wayt's hard work, relentlessness, and attention to detail to go for naught simply to satisfy an anticipated embargo.

My final bit of advice: Find someone with whom you can shoot the breeze—a professor, a scientist, a pundit, a colleague, a friend, a mentor. Exchanging ideas is a great way to keep you alert and to come up with fresh angles and perspectives. Good science journalism is, after all, less about having a science background than it is about having an inquisitive, tenacious mind.

2

Reporting From Science Journals

TOM SIEGFRIED

Tom Siegfried was born in Ohio and migrated to Texas, graduating from Texas Christian University in 1974 with majors in journalism, chemistry, and history. He earned a master's degree from the University of Texas in 1981. He was science editor at the *Dallas Morning News* from 1985 to 2004. He has written two popular science books: *The Bit and the Pendulum* (2000) and *Strange Matters* (2002). His work has been recognized with awards from the American Chemical Society, the American Psychiatric Association, the American Association for the Advancement of Science, and the National Association of Science Writers.

For police reporters, there are crimes. For political writers, elections. Sportswriters have games. And science writers have journals. In fact, there are more journal articles published every year than there are games, elections, and murders in all U.S. cities combined. So science writers must be selective. To select wisely, you'll need to know, first of all, what the major news-providing journals are, and what sorts of science they publish. You'll need to understand the different kinds of journals and different kinds of papers within them. And you'll need to comprehend how to navigate the elaborate web of censorship rules that most journals impose on reporters—a pernicious convenience known as the embargo system.

Once you know all that, you can concentrate on reporting and writing.

The Journal Menu

For science writers, the only journals of interest are those that are *peer-reviewed*, meaning that experts in the field have read the papers, and possibly suggested corrections and revisions, before the journal agreed to publish them. Traditionally, many science writers have focused on reporting from the "Big Four" peer-reviewed journals: *Science*, *Nature*, the *New England Journal of Medicine*, and the *Journal of the American Medical Association*.

Science and *Nature* are major sources of science news, and they should be. They are the premier interdisciplinary journals of the English-speaking world, and therefore ought to be publishing the most important research of the broadest interest to the scientific community. Naturally, such research is most likely to be of interest to the general public as well.

In recent years, the Big Four have been joined by several others as regular sources of science news—particularly the *Proceedings of the National Academy of Sciences*, the biology journal *Cell*, and the neuroscience journal *Neuron*. And the Nature publishing group has flooded the media journal market with a whole roster of specialty journals on such topics as neuroscience, biotechnology, genetics, and materials science. Other important journals for medicine include *Annals of Internal Medicine* and several published by the American Heart Association, such as *Circulation* and *Stroke*. An intriguing newcomer in late 2003 from the Public Library of Science is *PLoS Biology*, an "open-access" journal available free online.

The journals the media turn to most are not, however, the only sources of important scientific research, and for some fields they are not even the best. Depending on the scope of your beat, news will come to you from any number of other journals serving narrower segments of the scientific world.

In the physical sciences, for example, you will want to be familiar with the journals of the American Physical Society, including *Physical Review Letters* (publish.aps.org). The American Chemical Society (pubs.acs.org) also publishes a wide range of journals. For astronomy and astrophysics, you'll want to tune in to the *Astrophysical Journal*. For geology and the earth sciences, start with *Geology* and *Geophysical Research Letters*.

Many of these journals are available via the online service Eurekalert! (www.eurekalert.org), which posts "tip sheets" (restricted to registered journalists) announcing what the journals consider to be the best papers in each upcoming issue. Usually, full tables of contents are also available, as well as full text of the articles. *Nature* has its own press access Web portal. The American Physical Society offers an open Web page that alerts journalists to many upcoming stories (focus.aps.org), plus a restricted access site where reporters can acquire full-text papers.

So far, so good. But keep in mind that news also lurks in journals that don't advertise their existence. When you are reporting on a specific discipline, you should ask experts within that discipline which journals they regard as authoritative. When you identify good journals in a field, it's usually possible to sign up for e-mail alerts with tables of contents.

Another thing to keep in mind is that not all journals exist for the sole purpose of publishing original research. Many are devoted to "review" articles that help researchers keep up with new developments and trends in their fields. Usually, review articles are not a source of news, but they can provide important background for putting new reports in context.

Embargoes

A common feature of many major journals is their insistence on enforcing an *embargo* on release of their news. New papers (or drafts) are typically made available about a week before publication, with the understanding that reporters receiving this embargoed material agree to wait until the actual publication date to report it. Ostensibly this system gives reporters time to work on the story without fear of someone else's reporting it first.

If that were all the embargo system amounted to, it would not be so bad. But such journals usually also impose a gag order on authors of papers awaiting publication. In some cases, the scientists must sign a written agreement not to tell journalists about their work (except when the reporter has agreed not to violate the embargo). Some journals allow scientists to report their findings at scientific meetings, but not to answer journalists' questions about them. On occasion, journals have even prohibited scientists from presenting their work to other scientists at such meetings. You may freely report on what a scientist presents at a meeting you attend, of course, whether the journal likes it or not.

Be aware of how the embargo system operates and be alert to the possibility that someone else will in fact violate it. On major stories, it's a good idea to get your story done well in advance of the embargo date, so it will be ready to run right away if someone else breaks the embargo. Once any publication breaks an embargo, other media will no longer observe it.

Preparing

Thanks to the availability of journal papers in advance of publication, science writers usually have a fair amount of preparation time before applying fingers to keyboards. Take advantage. Don't wait until the last minute. Download the paper as soon as possible, and collect whatever peripheral information is available, such as news releases or commentary articles that accompany the paper.

You'll usually want to acquire additional background information from various sources. Google-search the authors to get some context about their research. Check your own publication's electronic morgue to determine what aspects of this research have already appeared. Do a Nexis search to find out what has been reported elsewhere. Check PubMed or other databases to find the authors' earlier papers and related papers by other scientists. If you're unfamiliar with the new paper's field, a general review article or a basic encyclopedia entry can familiarize you with essential terminology.

And then—and here is the key step in the process—*read the paper.*

Not all science journalists do. Some read the news release, glance at the paper, and then call up the researcher and ask a few questions. Go ahead and take that approach if your goal is mediocrity. If you want to be good, you have to learn how to read a scientific paper critically.

When I read a paper, I usually first scan the abstract and then read the introductory paragraphs to get a sense of the context for the research. I then go to the conclusions section at the end, so I'll know what the authors have to say about the ramifications of their work and what to pay attention to when reading the rest of the paper. Then I'll read the paper through, watching for things that might raise questions about the work (where did the data come from, how statistically significant are the results, any peculiarities about the methodology, presence or absence of control groups, etc.). Then I look at the data tables and graphs, trying to see if I can figure out how the data illustrate the conclusions the authors have stated.

All through this process, it's a good idea to jot down the questions that arise in your mind. The next step is deciding which scientists to pose them to.

Obviously, you need to talk with one of the authors of the paper. Typically, the first author listed is the person who did most of the work (often a graduate student or postdoc); the last named is the senior scientist or head of the lab (who often did none of the work). However, senior authors frequently have the best grasp of the research as a whole and are best able to answer questions and put it in context (and sometimes they actually did do a lot of the work).

Often it's a good idea to talk to more than one of the authors. They may have worked on different aspects of the study, and they may also have quite differing opinions on the meaning and significance of the results.

For most stories from journals, you'll need outside comment from sources not involved in the published paper. But some journalists (especially nonscience journalists) misunderstand this requirement. The point is *not* to find someone who disagrees with the results so that you can say that your story is "balanced." This is an idiotic idea, sometimes imposed by nonscience journalist editors with an archaic notion of "telling both sides" of the story. (This attitude is perhaps advisable when covering politics, or accusations of wrongdoing, but nonsense when applied to science. Otherwise every space story involving satellites would include a comment from the Flat Earth Society.)

In fact, the purpose of outside comment is to provide readers with an intelligent assessment from a knowledgeable specialist who is in a position to understand and appreciate the paper's significance.

It's important to realize, of course, that not all competent scientists would necessarily offer the same assessment of a given paper. You need to be aware, for example, if scientists in a given field are divided into camps with opposing views. In that case, it is perfectly appropriate to seek comment from members of each camp. It is irresponsible, on the other hand, to portray the views of a lone dissenter as equally meritorious to those reflecting an established scientific consensus.

You can find experts to call by checking the acknowledgments and the references at the end of the paper. You can ask the author of the paper to suggest people who are familiar with the work—and in fact, you can ask for names of people in the field who are likely to have a different (even disagreeing) perspective. Good scientists will tell you.

Another good approach, especially if you are in a hurry, is to identify a university or other institution that is prominent in the field. The public information officer there can usually put you in touch with an expert quickly. For a story involving subatomic particles, for instance, you might call Fermilab; for nanotechnology, you could call Caltech. Or you can call the press officer at the relevant scientific society—the American Astronomical Society for an astronomy story, for instance, or the American Geophysical Union for news in the earth sciences.

Checking the Facts

Don't Trust the Blurbs on Tip Sheets

They can be helpful, but they can also be wrong. (Just after writing this sentence, I received a tip sheet correction from *Nature.* Seems that the experiments on cat whiskers reported on the tip sheet were actually performed on rat whiskers.)

Don't Trust News Releases

They can be helpful, but they can also be wrong. Verify release information from the actual paper or the paper's authors (and whatever you do, don't lift quotes from the release). Double-check background information with other reliable sources.

Be Aware of the Pitfalls of the Peer-Review System

Some journals have more rigorous peer review than others, and even the best journals occasionally slip up. A paper once accepted for publication in *Physical Review Letters* purported to show evidence that the universe possessed a preferred direction of space. Now, anybody with an even elementary understanding of the universe knows that space is supposed to be the same in all directions. But here was a paper proclaiming that polarized radio waves preferentially twisted one way rather than another. When a paper expressing a claim of such magnitude gets published in such a prestigious journal, the claim warrants attention—and maybe even a story.

But I was not impressed. The study was based on a reanalysis of old data—observations not originally intended to test the space-direction issue. The statistical significance of the result was borderline. And some of the data that didn't support the conclusion had been thrown out. I decided not to write a daily story.

Some other newspapers did run the story. Within a week or so, though, papers by other physicists began appearing on the Internet, rebutting the *Physical Review Letters* paper's conclusions. The paper was quietly forgotten. It was a nonstory, one of many published papers of no lasting (or even temporary) significance—even though it came wrapped in all the trappings of the real stories that science journalists are supposed to write. The lesson is simple: Just because a paper gets published in a peer-reviewed journal, that doesn't mean it warrants a story.

Ask a Paper's Authors About Previous News Coverage of Their Work

You want to make sure that what you think is new really is, and wasn't widely reported last year after a presentation at a meeting.

Ask About Potential Conflicts of Interest

For example, do any of the researchers have a financial stake in a company that could profit from a study's findings? (But be careful in reporting such conflicts—a financial interest does not automatically invalidate the results of a properly conducted study. You have to judge whether stating the conflict might be misinterpreted as calling the research into question.)

Check Trivial Facts

For example, check a scientist's affiliation and title. Sometimes the title page of a journal article contains mistakes on such matters.

Writing the Story

From the moment you begin considering a story on a journal paper, you should be thinking about the story's opening sentence or paragraph: the all-important *lead* (or *lede*, as it's commonly spelled in our world). What is the key new point? What is the most important, most interesting thing about it? How can you capture all that in a concise, clear, and catchy way?

From then on, it's go with the flow. Support the lead with the facts. Provide a quote that dramatically expresses significance. Work in the background that provides context—both basic information and previous relevant findings. Give details that answer all the questions you can imagine a reader asking. And say what will or will not happen next. Sometimes you also need to tell what the results do *not* mean, as in medical stories where a promising finding does not imply an immediate cure.

But always remember, sometimes the best thing to do is not to write a story at all. Daily stories from journals are a staple of science journalism, but they are far from all that science journalism should be. It's often wiser to wait for scientists to publish more research or for you to do more reporting. Ultimately, you serve your readers best when you write stories that report the work of science with context and perspective.

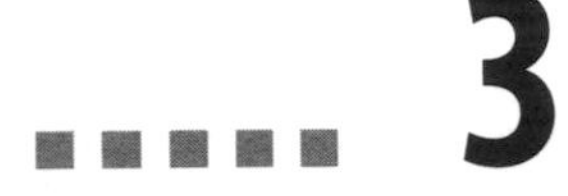

Understanding and Using Statistics

LEWIS COPE

Lewis Cope was a science writer at the *Minneapolis–St. Paul Star Tribune* for 29 years. He is a member of the board of the Council for the Advancement of Science Writing and a former president of the National Association of Science Writers. He is co-author (with the late Victor Cohn of the *Washington Post*) of the second edition of *News & Numbers: A Guide to Reporting Statistical Claims and Controversies in Health and Other Fields* (1989, 2001).

A doctor reports a "promising" new treatment. Is the claim believable, or is it based on biased or other questionable data? An environmentalist says a waste dump causes cancer, but an industrialist indignantly denies this. Who's right?

Meanwhile, experts keep changing their minds about what we should eat to help us stay healthy. Other experts still debate what did in the dinosaurs. Which scientific studies should you believe?

This chapter deals with the use (and sometimes misuse) of statistics. But don't let this S-word panic you. Being a good science writer doesn't require heavy-lifting math. It does require some healthy skepticism, and the ability to ask good questions about various things that can affect research studies and other claims. To separate the probable truth from the probable trash, you need to get answers to these questions:

1. Has a study been done, or is a claim being made on the basis of only limited observations? If a study was done, how was it designed and conducted?
2. What are the numbers? Was the study large enough (did it have enough patients or experiments or whatever) to reach believable conclusions? Are the results *statistically significant*? That phrase simply means that, based on scientific standards, the statistical results are unlikely to be due to chance alone.
3. Are there other possible explanations for the study's conclusions?
4. Could any form of bias have affected the study's conclusions, unintentional or otherwise?
5. Have the findings been checked by other experts? And how do the findings fit with other research knowledge and beliefs?

Principles for Probing Research

To find the answers to these questions, we must understand five principles of scientific analysis:

1. The Certainty of Some Uncertainty

Experts keep changing their minds not only about what we should eat to stay healthy but also about what we should do when we get sick. A growing number of drugs and other treatments have been discredited after new research has raised questions about their effectiveness or safety. Even the shape of the universe (more precisely, how astronomers *think* it's shaped) has changed from one study to another.

To some, these and other flip-flops give science a bad name. But this is just part of the normal scientific process, working as it's supposed to work.

Science looks at the statistical probability of what's true. Conclusions are based on strong evidence, without waiting for an elusive proof positive. The complexities of nature and the research process can add to uncertainty.

But science can afford to move ahead because it is always an evolving story, a continuing journey that allows for mid-course corrections. In fields from medicine to astronomy, from geology to psychology, old conclusions are continuously being retested—and modified (or occasionally abandoned) if necessary.

We need to explain this to our editors and news directors, and to our readers and viewers. Some uncertainty need not impede crucial action if the public understands why *at best* almost all a scientist can say is: "Here's our strong evidence that such-and-such is probably true. Please stay tuned as we work to learn more."

As we move into the details, keep in mind that not all research is equal.

2. *Probability, Power, and Large Numbers*

Have you heard the one about a new drug tried in mice? "Thirty-three percent were cured, 33 percent died—and the third mouse got away." This old joke reminds us how important numbers are in assessing the worth of a study.

The more patients in a study, the better. The higher the success rate with a new treatment, the better. The more weather observations that meteorologists make, the better they can predict whether it will rain next week. Here's how the numbers affect the statistical *probability* that something is true:

A commonly accepted numerical expression is the *P* (probability) value, determined by a formula that considers the number of patients or events being compared. A *P* value of .05 or less is usually considered statistically significant. It means that there are 5 (or fewer) chances in 100 that the results could be due to chance alone. The lower the *P* value, the lower the odds that chance alone could be responsible.

Put another way: The larger the number of patients (or whatever), the more reliable the *P* value.

There are two related concepts. This first is called *power*. This is the likelihood of finding something if it is there—for example, an increase in cancer cases among workers exposed to a suspect chemical. The greater the number of observations or people studied, the greater the power to find an effect. A new drug's risk of causing a rare but dangerous side effect may not become clear until it has been marketed and then used by many tens of thousands, sometimes even millions, of patients.

The second is called *statistical strength*. If a pollutant appears to be causing a 10 percent increase in illnesses above background levels, it may or may not turn out to be a meaningful association. If the risk is 10 times greater (like the relative risk of lung cancer in cigarette smokers versus nonsmokers), the odds are very strong that something is happening.

Science writers don't have to do the math. They just have to ask researchers: *Show me your numbers.*

Key questions to pose: *Are all your conclusions based on statistically significant findings?* (Be leery if they aren't, and warn your readers or viewers.) *What are the P values—the chances that key findings are due to chance alone? Was your study big enough to find an effect if it was there? Are there other statistical reasons to question your conclusions? Are larger studies now planned?*

But just because findings are statistically significant, and have sufficient power, and so forth, doesn't mean that the findings are necessarily correct, or important. So our list continues.

3. *Is There Another Explanation?*

Association alone doesn't prove cause and effect. The rooster's crowing doesn't cause the sun to rise. A virus found in patients' bodies may be an innocent bystander, rather than the cause of their illness. A chemical in a town's water supply may not be the cause of illnesses there. Laboratory and other detailed studies are needed to make such cause-and-effect links.

A case history: A few scientists (and many news reports) have speculated about whether childhood immunizations might be triggering many cases of autism. But most experts believe this is coincidence, not causation. The "link" is only that autism tends to start at the same age that children get a lot of their shots, these experts say. The concern now: Some worried parents may delay having their children immunized against measles and other dangerous diseases because of a false fear about autism. In many press reports, the missing numbers are the tolls these childhood diseases took before vaccines were available.

A study's time span can be very important. Climate studies must look at data over many years, so they won't be confused by normal cycles in the weather. A treatment may put a cancer patient into remission, but only time can tell if this provides a cure or even lengthens survival.

Some patients may drop out during the course of a long study. If they leave because they aren't doing well, this may confuse the study's numbers.

Then there's the *healthy worker effect.* A researcher studies workers who have been exposed to a risky chemical and finds that, on average, they are even healthier than the general population. But don't absolve the chemical yet. Workers tend to be relatively healthy—they have to be to get and then keep their jobs.

And expect some normal variations. People are complex. There may be day-to-day biological variations in the same person—even more between populations. Similar studies may have small differences in results, occasionally even marked differences, due to variability or other research limits.

The list could continue, but broad questioning may keep you from going astray. Ask the researcher (and ask yourself as well): *Can you think of any alternate explanations for the study's numbers and conclusions? Did the study last long enough to support its conclusions?*

In science, the term *bias* is used to cover a wide range of failures to consider alternate explanations. But science writers also need to probe the possibility of another type of bias by asking researchers: *Who financed your study?* Many honest researchers are funded by companies with an interest in what's being studied. You should ask about such links, and then tell your readers and viewers about them.

4. *The Hierarchy of Studies*

For costs and other reasons, not all studies are created equal. As a result, you can put more confidence in some types of studies than in others.

In biomedical research, laboratory and animal studies (even those with many more than three mice) should be viewed with particular caution. But they can provide vital leads for human studies.

Many epidemiological and medical studies are *retrospective*, looking back in time at old records or statistics or memories. This method is often necessary but too often unreliable, because memories fade and records frequently are incomplete. Far better is a *prospective* study that follows a selected population for the long term, sometimes decades.

The "gold standard" for clinical (patient) research is a *double-blind* study, with patients randomly assigned to either a treatment group or a *control* (comparison) group. The patients in the control group typically receive placebos. The blinding (where practical) means that neither the researchers nor the patients know who has been assigned to which group until the study is completed. This keeps expectations and hopes from coloring reported results. Patients are randomly assigned to the two groups so that a researcher won't subconsciously put a patient who's likely to do better into the treatment group.

Less rigorous studies still may be important—sometimes even necessary. But put more faith in more rigorous studies.

Ask researchers in all scientific fields: *Why did you design your study the way that you did? What cautions should people have in viewing your conclusions?* And often: *Is a more definitive study now needed?*

5. *The Power of Peer Review*

You can give a big plus to studies that appear in *peer-reviewed* journals, which means these studies have passed review by other experts. But this is no guarantee. Reviewers are human. Good stories can also come out of science meetings before they appear in peer-reviewed journals, and even from scientists who are just beginning studies. But these research stories demand more cautious reporting, more checking with other experts.

Ask researchers: *Who disagrees with you? Why? How do your findings and conclusions fit with other scientific studies and knowledge?*

The burden of proof rests with researchers seeking to change scientific dogma. And always, science loves confirming studies. Science writers should look for consensus among the best studies.

In *News & Numbers*, we give this bottom-line advice: "Wise reporters often use words like 'may' and 'evidence indicates,' and seldom use words like 'proof.'" Spell out the degree of uncertainty involved in what you are reporting. Provide appropriate cautions and caveats for added credibility.

Dollars and Averages

Ask about costs. It's fun to write or talk about a futuristic scheme to move some asteroid away from a possible collision course with Earth, but *how much would it cost? Can we afford it?* The public particularly wants to know the price tag for any new medical treatments. Ask, *Will it be so expensive that it's unlikely to see widespread use?* If the researchers don't have cost estimates, that's news, too.

Don't be misled by averages. People can drown in a lake with an average depth of four feet when it's nine feet deep in the middle. The average person in a study exercises three hours a week; not mentioned is that most of the participants don't exercise at all, while a few are zealots. Ask, *What are the numbers behind the average?* A radio report said that "*you'll* live longer" if you exercise and eat a prudent diet. The evidence is only that people will live longer *on average*. "You" only increase your chances of doing so.

Rates and Risks

Avoid rate confusion. The *Washington Post* ran an article with the headline "Airline Accident Rate Is Highest in 13 Years." The story, like many others that misuse the term "rate," reported no rate at all, merely death and crash totals. A correction had to be printed pointing out that the number of accidents per 100,000 departures—the actual rate, the "so many per so many"—had been declining year after year. (The headline would have been technically correct if it had simply said "Airline Accidents Highest in 13 Years." But in this and many other cases, I think that the rate is the fairest way to judge what's really happening.)

Watch risk numbers. Someone cites deaths per ton of some substance released into the air, or deaths per 10,000 people exposed. Someone else cites annual deaths, or a 10-year death total. There are many choices to make something seem better or worse. Make sure you get a full, fair picture.

While you're at it, pay attention to the difference between *relative risk* and *absolute risk*. Relative risk is a measure of the increased risk of developing an illness or disorder. Example: A study concludes that people exposed to a chemical (say a hypothetical Agent Purple) are twice as likely to develop a particular cancer as the people who were not exposed to that chemical. The relative risk is 2.

But in total lives affected, even a large increased risk for a rare illness is not as important as a small increased risk for a common illness. Absolute risk takes this into consideration. It calculates the "number of cases per X thousand population per year." Relative-risk calculations can be important in discovering a threat; absolute-risk calculations can be useful to show the public health or clinical impact.

View clusters with caution. When you hear of a very high number of cancer cases clustered in a neighborhood or town, more study may be warranted, but not panic. With so many communities across our nation, by chance alone a few will have many more than their share of cancer cases (or birth defects or whatever). This is the Law of Small Probabilities.

Put the burden where it belongs. Someone says, "How do they know this stuff isn't causing harm?" Science can't prove a negative. The burden of providing at least some evidence is on the person making a claim of harm.

Potential Perils in Polling

Polls go beyond politics. They can help us learn what people do (and don't do) to stay healthy, and whether the public thinks we should spend more on space exploration or whatever. But to be credible, polls must pass scientific analysis.

The people interviewed must be a *random sample* of the population whose views we want to learn about (for instance, registered voters in the Midwest, or teenage smokers). Caution: TV talk shows often ask people to phone in their poll answers. But only the show's viewers will know to call, and only those with strong views are likely to call. That's not a random sample, and it's not a scientific poll.

The more people interviewed in a poll, the smaller the *margin of sampling error*. This margin of error may be, for example, "3 percentage points, plus or minus." That means that in 19 out of every 20 cases (the statistically significant standard), the poll's results will be accurate to within 3 percentage points—*if* all else is right with the poll.

The poll's questions must be crafted to eliminate any bias that might nudge those polled to answer in a particular way. Ask, *What's the exact wording of the questions? Who paid for this poll?*

And polls are only a snapshot of what people say at a particular time. This may change.

The bottom line for polls and all sorts of scientific studies: Look at the numbers, keeping in mind that bigger tends to be better. Ask yourself if there are any alternate explanations for the poll's, or study's, conclusions. Consider any possible biases, intentional or otherwise. And keep in mind the certainty of some uncertainty.

This chapter is based on concepts covered in detail in *News & Numbers: A Guide to Reporting Statistical Claims and Controversies in Health and Other Fields* (Iowa State Press; 1st ed., by Victor Cohn, 1989; 2nd ed., by Victor Cohn and Lewis Cope, 2001).

4

Writing Well About Science: Techniques From Teachers of Science Writing

While the rest of this book deals with how to handle a variety of jobs, cover certain topics, and use specific tools to get the job done, this chapter opens the classrooms of some teachers of science writing. Their techniques help correct common problems and reveal strategies for writing clearly and beautifully.

Deborah Blum is a professor of journalism at the University of Wisconsin–Madison. Mary Knudson teaches science writing in Johns Hopkins University's Master of Arts in Writing Program in Washington, D.C. Ruth Levy Guyer teaches in the same Hopkins program, as well as at Haverford College in Pennsylvania, and at the UCLA School of Medicine. Sharon Dunwoody is the Evjue Bascom Professor of Journalism and Mass Communication at the University of Wisconsin–Madison. Ann Finkbeiner runs the graduate program in science writing at the Writing Seminars at Johns Hopkins in Baltimore. And John Wilkes is the director of the Science Writing Program at UC–Santa Cruz.

Ten Time-Tested Tips

1. *Read your work out loud.* You will be able to hear rhythm and flow of language this way, and you really cannot hear it when reading silently.
2. *Don't be shy.* Ask other writers to read a draft for you. Everyone gets too close to the story to see the glitches, and a dispassionate reader is a writer's best friend. Good writers gather readers around

them for everything from newspaper stories to whole books (which require really good friends).

3. *Think of your lead as seduction.* How are you going to get this wary, perhaps uninterested reader, upstairs to see your etchings? You need to begin your story in a way that pulls the reader in. My favorite basic approach goes seductive lead, so-what section (why am I reading this), map section (here are the main points that will follow in this story). That approach leads me to my next tip, which is
4. *Have a clear sense of your story and its structure before you begin writing.* If you think of a story as an arc, in the shape of a rainbow, then it's helpful to know where it will begin and where it will end so that you know in advance how to build that arc.
5. *Use transitions.* A story has to flow. Leaping from place to place like a waterstrider on a pond will not make your prose easy to follow.
6. *Use analogies.* They are a beautiful way to make science vivid and real—as long as you don't overuse them.
7. *In fact, don't overwrite at all.* And never, never, never use clichés. If you want to write in your voice, generic language will not do. In my class, there are no silver linings, no cats let out of bags, no nights as black as pitch. A student who uses three clichés in a story gets an automatic C from me.
8. *Write in English.* This applies not only to science writing but to all beats in which a good story can easily sink in a sea of jargon.
9. *Picture your reader.* I find it helpful to imagine a specific reader who is unnerved by science to begin with and would stop reading my story the minute I threw a multisyllabic medical term in her face. Yes, her face. My reader is an elderly woman with curlers in hair, half-dozing over the paper. If I can snare her, the science-savvy reader is a snap.
10. *Have fun.* Science is intriguing, funny, and essential to everyday life. If we write too loftily, we lose some of the best stories and the ones that our readers most relate to.

DEBORAH BLUM

On Explaining Science

1. The question is not "should" you explain a concept or process, but "how" can you do so in a way that is clear and so readable that it is simply part of the story?

2. Use explanatory strategies such as . . .
    - Active-voice verbs
    - Analogies and metaphors
    - Backing into an explanation, that is, explaining before labeling
    - Selecting critical features of a process and being willing to set aside the others, as too much explanatory detail will hurt rather than help.
3. People who study what makes an explanation successful have found that while giving examples is helpful, giving *nonexamples* is even better.

    Nonexamples are examples of what something is *not*. Often, that kind of example will help clarify what the thing *is*. If you were trying to explain groundwater, for instance, you might say that, while the term seems to suggest an actual body of water, such as a lake or an underground river, that would be an inaccurate image. Groundwater is not a body of water in the traditional sense; rather, as Katherine Rowan, communications professor, points out, it is water moving slowly but relentlessly through cracks and crevices in the ground below us.

    Or if you were trying to explain what a good survey is, you might reflect on the importance of a representative sample, a good response rate, and so on, but then offer examples of what a good survey is *not*: stopping people at the local mall (haphazard sample), asking magazine subscribers to return a survey insert in their magazine (lousy response rate), and so forth. This gives your reader a different way of looking at a new idea.
4. Be acutely aware of your readers' beliefs. You might write that chance is the best explanation of a disease cluster; but this could be counterproductive if your readers reject chance as an explanation for anything. If you are aware that readers' beliefs may collide with an explanation you give, you may be able to write in a way that doesn't cause these readers to block their minds to the science you explain.

SHARON DUNWOODY

On Writing Clearly and Logically

In fiction where events are chosen for their emotional tone—a character disappears, a word is misspoken—any number of events will suit. Nor do two neighboring events need any connection other than emotion: Chronology doesn't much matter; logic certainly doesn't. Just add that scene with the misspoken word to the scene where the character disappears into the fog, and the reader knows exactly what happened. In fiction, connections don't have to be explicit; we get it.

In science writing, where events proceed on logic, order counts. In each paragraph, the sentences have a right order; and in each sentence, the words have a right order; and all you have to do is find the order. A Sid Harris cartoon shows Johannes Kepler telling his colleagues, "So you see, the orbits of the planets are elliptical," and the colleagues thinking one by one, "What's an orbit?" "What's a planet?" "What's elliptical?" Those colleagues' biggest problem isn't the jargon; it's that Dr. Kepler obviously didn't present his case in the right order.

Sometimes when the order doesn't seem right, it really is; and what's wrong is that some of the logic is missing. The easiest examples of missing logic are on the small scale. "Because the speed of light is constant, we see galaxies that are distant in space also distant in time." That sentence is grammatically respectable and factually accurate, but for the lay reader, it's close to nonsense. The problem is that the logic connecting light, the galaxy in space, and the galaxy in time is missing. The reader needs to be held by the hand, and walked through the idea, every step, step by step. One way of doing this is a brilliant but obvious rule that I made up: Begin each sentence with the word or phrase that ended the previous sentence. So: "The only way we see galaxies is by their light. Light leaving a galaxy at a certain distance and traveling a fixed speed takes, say, 100 years before we see it. We see the image of the galaxy as it was 100 years ago." The pattern is A–B, B–C, C–D.

The AB/BC rule also works on the mid-scale, with paragraphs. In fact, the rule is just another way of making transitions. If each paragraph is a single idea—as Strunk and White rightly assert it should be—then transition sentences provide the connections between ideas. Transition sentences tell the reader why, having read that, he is about to read this. If the transition to a new paragraph seems awkward—"But first, some background"—the ideas are probably in the wrong order and you've got structural trouble. Another symptom of structural trouble is the repetition of subjects in an article: "As pointed out earlier," or "Getting back to the early universe," or "We will come back to this subject in a later section." When you discover these symptoms, return to the AB/BC rule, take a deep breath, and start over.

AB/BC also works nicely with Strunk and White's most unbreakable rule: The most important word in a sentence comes at its end. I happened to be interviewing the psychologist Marvin Minsky once and asked him about that sentence's-end rule: Naturally, he said, you remember best what you heard last. That rule also applies to paragraphs: The most important sentence comes at the end. The reason the rule is so good is that by the time you've finished writing a paragraph full of sentences ending with their most important words, you probably can figure out what the paragraph was about. And by the time you've written a story full of paragraphs ending with their most important sentences, you probably know what the article was about.

ANN FINKBEINER

■ ■ ■

Think of the path of logic as comparable to the alphabet. In order to recite the alphabet correctly, one must begin at A and go logically to B and then to C and so on to Z. Not one letter can be skipped in the alphabet, and not one step can be overlooked in the path of logic pursued through the story. The writer should envision a smart reader who is unfamiliar with and uninformed about the subject, but not stupid. That reader can learn anything she needs to know as long as the writer writes what the reader needs to know when she needs to know it. Make no leaps of knowledge or faith. The task of explaining something new to this attentive reader is straightforward, precise, interesting, and extremely challenging.

RUTH LEVY GUYER

■ ■ ■

Deletable phrases: "There are," "it is," and so on. "There" and "it" are pronouns that refer to nothing and only take up space. The sentence accordingly loses zip: "There are 10 billion neurons in the brain" versus "The brain has 10 billion neurons."

Deletable words: Excess "the's" and all "very's." "Very" is fine in spoken English but is counterproductive in written: "She is very beautiful" versus "She is beautiful." Ration adverbs strictly.

ANN FINKBEINER

Storytelling

Science is a process rather than a product, and this is why it lends itself to storytelling. Scientific discoveries are made by people; they don't just happen. Good writers give readers a picture of scientists carrying out experiments, recording cause-and-effect relations, documenting observations, disturbing steady states, and being excited and sometimes startled by their findings. Authentic scientists expect the unexpected, and when it happens they love it.

Explaining the general, broader significance of a discovery is also crucial. There may be little that is absolutely new. Nature is said to be parsimonious. If something works one time, nature uses it elsewhere.

A thoughtful writer will dig deep into his or her own interests, strengths, biases, and agendas and not only develop the story itself but also tie it to other things in the world—in science and also in the broader literature and culture—that add interest and insight to the story. The writer who attends closely to both deep and broad issues is the one who will create something that is different from what other writers are producing. This writer will write the story that is worth the readers' time.

RUTH LEVY GUYER

■ ■ ■

Set a pace. Once you've drawn readers in, you want them to be able to read quickly through your story. If you can read a story quickly, it means the story was well written. A well-written story has a good pace—at times leisurely, describing scenes, building anticipation; at times quickened, revealing action, terse dialogue. The pace of the story is what will keep your readers reading to the end. If you don't set a pace and sustain it through the last graph, you won't have very many readers reading that last graph.

So, how do you set a pace? Frankly, you have to play with it a lot. But you can start with a framework. Use active voice and powerful verbs. Use present tense to create immediacy and adventure. Use past tense if immediacy is not needed. Put short sentences in strategic places as segues to the next segment of your story. Alliteration, using examples in sets of three, and varying your sentence structure all help to create a rhythm. And, very important, eliminate clutter.

Narrative writing is essentially a combination of fiction techniques that are very useful in telling medical and science stories. Even if you are not writing a narrative, think of yourself as a storyteller. Use narrative writing for an entire piece or only a portion of a story. Here are the basics: details, anticipation, quotes.

Details give such vivid descriptions that you reach out and put the reader smack into the story. Anticipation builds interest in reading on by giving a hint of what is to come. Quotes bring your story to life, are authoritative, raise provocative questions. Quotes are heartbreaking, whimsical, funny. Quotes make the complex understandable. Quotes give the other side of the story. Conversational quotes help set the rhythm of a narrative. If you put a quote in a story that does none of these things, strongly consider taking it out. Your interviews will bring lots of quotes that are verbose, empty, and loaded with jargon.

MARY KNUDSON

■ ■ ■

The three main writing problems manifested by most of my scientist-students, and my tips for fixing them:

1. Years in academia, where one is rewarded according to the fine distinctions and logical complexity of one's thought, have bred some bad habits in many of my students. They tend to write too long, and to insist on a degree of thoroughness and a level of detail for which a general reader has little patience.

   Also—and this applies to all students of writing, not just scientists—I'm convinced that computers encourage syntactic looseness and a general logorrhea, especially now that one can publish on the Web, where most copy doesn't need to be fit to a prescribed space. First drafts of stories can contain many sins, and probably should, but those sins should be expiated, and lots of water should be wrung out, before we impose a piece of writing on a reader. So, at Santa Cruz we focus on the many ways to find the most direct path to a clear statement of an idea.

   Tip: Edit yourself to make your points as economically, yet memorably, as possible.
2. My students tend to overresearch stories, then to overexplain the technical elements they've laboriously mastered. The editor of the *Santa Cruz Sentinel*, the local daily newspaper, uses my students as interns. He once told me, "They're very smart, but if you ask them what time it is, they tell you how the watch works."

   Tip: Don't get lost in the forest. Apportion the number of words you give to a story element according to the importance of the element to the story, not to the element's technical complexity.
3. My students can be too respectful of big-name scientists—indeed, of all scientists. This can weaken their interviewing and writing. The students are fresh from the lab and, understandably, still want to be seen as scientists. Researchers, of course, are delighted to be interviewed by a scientist rather than a typical general assignment reporter, and are all too happy to treat her or him as a colleague rather than a reporter. I tell my students to resist the

deep-rooted, often unconscious desire to be approved of by sources. I tell them to imitate *New Scientist*'s cheeky approach to scientists and science.

Tip: Be cheeky.

In addition to these tips, I'd like to quote *New Yorker* writer Ved Mehta, who said he learned from editor William Shawn to work always toward attaining "clarity, harmony, truth and unfailing courtesy to the reader."

JOHN WILKES

▪▪▪▪▪ 5

Taking Your Story to the Next Level

NANCY SHUTE

In 20 years as a journalist, Nancy Shute has reported firsthand on the outbreak of SARS, detailed forensic scientists' attempts to identify the remains of the victims of 9/11, and showed how cartoonist Gary Larson employed science to hilarious effect. Her work has taken her from the Russian Far East, where she became the first Fulbright Scholar in Kamchatka in 1991, to the Galapagos and Iceland. Her work has appeared in publications from *Smithsonian* and *Outside* to the *New York Times.* She has served as assistant managing editor at *U.S. News & World Report,* directing the magazine's science coverage, and is now a senior writer at the magazine, based in Washington, D.C.

"Don't pick the hard stories, sweetheart," an editor told me long, long ago. "Those are the ones that will break your heart."

Nonsense, I thought. I was young and ambitious and eager to chase a story through multiple all-nighters. He was old and wily and appreciated those stories that would glide through the copy desk and get him home in time for a glass of scotch and dinner with the family.

Now, more than 20 years after getting that good advice, I too appreciate the easy stories. But I'm still trying for the hard ones. Every few years, if I'm lucky, I manage to pull one off. When I do, the small, secret joy of having done so sustains me through months of too-short deadlines and too-tight space.

In thinking about what elevates a story from okay to prizewinner, from another day at the office to the top of the clip file, I think again about that long-ago editor, a grizzled veteran of the *Saturday Evening Post*. Don't try to be different, he said. Write about what everyone else is writing about. Those are the big stories, the ones that matter. And he was right. In covering science and medicine, we're blessed with big stories galore. Cloning, cancer, Mars exploration, anthrax, the Big Bang, climate change, nanotechnology, heart disease it's birth, death, creation, the meaning of life. If that can't get you on page A1, what can? But that very abundance, and the flood of data that bears those stories along, make it all too tempting to settle for the easy get—to write off the journals, take your lead from the *New York Times*, and get by. A great story demands more.

I like to think of journalism as bricklaying—a noble craft, but a craft all the same. To build a wall, I need bricks. To build a noble wall, I need the best bricks ever. Facts are the bricks of a story, and finding the right bricks requires serious reporting. I can't say that exhaustive research and reporting will guarantee a great story, but I've never been able to pull one off without it. The process starts so innocently. I notice that I've started to clip articles on a single subject, one that I've never particularly cared for before—say, population genetics. Suddenly I find myself fascinated by the new ability to use DNA testing to trace the movements of early humans out of Africa, into Europe, across Asia. I need to create a population genetics folder in my e-mail software to collect all the journal citations and random e-mails that I'm accumulating on the subject. I start angling for a little assignment that will let me spend a week at a Cold Spring Harbor meeting on genetics. And I read and read. Soon my desk and office floor are piled with journal articles, notebooks, and photocopied book chapters. You know these telltale signs. I'm overreporting.

What a waste, you say; be efficient, cut to the chase. But I know that in the process of reporting, I'm educating myself. For a technically complex story like population genetics, I need extra time to grasp haplotypes—groups of genetic variations that can be used to track kinship or the lack thereof—and genetic drift (forgive me, I was an English major). At a newsweekly, we don't have the luxury of running story after story on the same subject, building sources and knowledge along the way. Space is too tight. So I'll wedge the extra reporting in among shorter assignments, freelance a tangential piece that will get me to the right researcher or the right town, stay at the La Quinta on my own dime. Whatever it takes to buy the time I need to understand this story.

Start looking closely at the stories that grab you, and you'll see the sweat stains. Consider Kyla Dunn's 2002 *Atlantic Monthly* story "Cloning Trevor." She picked a huge subject: human cloning. Thousands of stories have been written about cloning, and hundreds of those have been about Advanced Cell Technology Inc., a struggling biotechnology company that is attempting to create cloned human embryos for stem cell therapy. What made Dunn's story worthy of the 2003 NASW Science-in-Society Award was her doggedness in documenting the personal, political, and scientific difficulties besetting human cloning.

Dunn put in the time. She spent weeks watching ACT researchers attempt to create a cloned embryo using skin cells from Trevor Ross, a young boy with X-linked adrenoleukodystrophy, a rare, often fatal genetic disorder. She talked with Trevor's parents, waded through NIH reports and congressional testimony, read article after article on the cloning debate. But sweat alone, alas, isn't enough to transport a story from the standard to the sublime. Dunn not only knew that she had to report; she knew that she had to edit. We don't read about how frustrating those months of reporting must have been for her, waiting for the ACT folks to come up with viable donor eggs. Instead, we are suddenly in the lab late on a January night. The researchers are attempting to fuse one of Trevor's cells with a rare donated human egg cell. "Using a tiny glass rod shaped like a miniature fencing foil, with a bulbous tip that can prod without piercing, [Vice President for Research Jose] Cibelli positioned the egg between the two electrodes at the bottom of the dish. His goal was to line it up so that the wires would send maximum current directly through the two cells, pushing them toward each other and confusing their membranes enough to make them fuse. Too little current and the cells wouldn't fuse, too much and the egg would be 'fried,' so to speak." Cibelli flips the switch. The fusion fails. "It was midnight, and the dispirited team began breaking down the equipment. 'We'll have more eggs,' Cibelli said, to nobody in particular. 'Hopefully, anyway.'"

In truth, it wasn't Dunn's description of the Bush administration's position on cloning that got me. It was poor Jose Cibelli, frying $20,000 worth of donor eggs at midnight in his mad-scientist quest to be the first to master therapeutic cloning. Indeed, finding the right people can turn a by-the-numbers story into a poem. Anne Fadiman's 1997 book, *The Spirit Catches You and You Fall Down*, could be summarized as "cross-cultural misunderstandings in medicine can kill you." But Fadiman, following that dry creek bed, moved to Merced, California, eased her way into the Hmong immigrant community there, and waited and listened until she found the right people. Thus, she was able to tell the heartbreaking tale of Lia Lee, a young Hmong girl with severe epilepsy. Fadiman showed how the wildly divergent views of health and illness, body and spirit held by her devoted parents and her conscientious doctors doomed Lia to a living death.

As I reported my 2001 story on population genetics, I discovered that not only were researchers running their own DNA to check their haplotypes, just for fun, but a few also were running DNA tests for individuals curious about their own heritage. Other geneticists were deeply distressed by this do-it-yourself approach, concerned that laypeople could misunderstand information about inherited disorders, or that someone might discover that the man they called Dad really wasn't Dad at all. Amidst all that, I saw an amazing tale emerging on the swift democratization of new technology. I had to find those genetic pioneers. It was tough work, since researchers are barred for privacy reasons from revealing their clients' names and don't relish spending time shuttling between patient and reporter getting consent for interviews. So I haunted genealogy Internet chat rooms, pestered population geneticists to spread the word, talked to lots of anthropologists, even asked workers in genetics labs if they had run their own DNA and if they would talk about it.

Thus, I found Andy Carvin, a 29-year-old Internet policy analyst who used his own DNA to probe his family roots in Ukraine. Carvin had sent his DNA to Family Tree DNA, a small Houston firm created for just such quests, because his father had often said that they were cohanim, members of an ancient Jewish priestly caste. New research tracing male Y chromosomes had tracked genes shared by cohanim back 3,000 years, to the time of Aaron, the Biblical progenitor of the caste. Through genetic testing, Carvin found out not only that his Y chromosome had the cohanim markers, but that his markers matched those of another man in the database, making it likely that they shared a forefather within the past 250 years. So in November 2000, Carvin took the train to Philadelphia and met 59-year-old Bill Skwersky, his genetic cousin. "We immediately hit it off," Carvin said. "I felt like I was visiting one of my uncles."

Another genetic pioneer I found was Pearl Duncan, a Jamaica-born writer who asked Ghanaian churchgoers in New York to give her cheek-swab samples of their DNA in an effort to help her reconstruct family history lost on the slave ships from Africa. She then had a private lab compare the churchgoers' DNA with her father's, and found enough of a match to convince her that the Ghanaian nicknames among her ancestors were no fluke. And Doug Mumma, who searched the Internet for people with similar surnames, paid for Y-chromosome testing of strangers in Germany, and found relatives there. I searched long and hard until I found an Alaskan Aleut elder who not only had had her DNA tested, but who was willing to talk about it. In fact, she was amused by her Siberian roots. My search took way too much time, and many days along the way I doubted if the people I was looking for even existed. But when I finally sat down to write, I knew I had something good.

The fourth hallmark of a great story, aside from a big subject, obsessive reporting, and the instinct to find the right people, is perspective. Readers need to know the broader context, why this story matters and how the science fits into human history. It's particularly important now, when most people get their news from electronic media where perspective is in woefully short supply. In her book, Fadiman had the luxury of space, and was able take several chapters to show how the Hmongs' flight from persecution in Laos and long stays in refugee camps made their eventual settlement in the United States so problematic. In writing newsweekly articles, I'm lucky if I have space for a few paragraphs to describe the past 20 years of genetic research, from the 1983 discovery of the PCR process that made it easy to duplicate and study DNA, to recent work showing that Neanderthals were almost certainly not human ancestors. But those few paragraphs may be the most important ones. They show readers not only how the science has developed—what's worked and what hasn't—but also how the changing attitudes of the public, the imperatives of culture, change the science. The world makes more sense, if only for a bit.

I'm still fond of a story I wrote back in 1997 about the struggles of researchers who were attempting to help treat the 20 percent of Americans who are problem drinkers. Europe, Great Britain, and Australia have long defined problem drinking as a public health issue and have used harm reduction, a strategy that focuses on practical efforts to minimize the damage caused by substance abuse, rather than seeking abstinence. But researchers who have successfully done similar work in the United States, notably in getting college students and young military recruits to moderate their drinking, have been vilified by a treatment community focused on abstinence as the only answer. It was only after I studied America's long, ambivalent romance with alcohol, and learned that for the past 100 years alcohol abuse has been addressed largely as a moral failing, despite mounting evidence for the genetic factors involved, that I was able to explain to readers why moderate drinking is even today almost never offered as a treatment option.

When a good story idea, meticulous reporting, great characters, and the right perspective combine in a single article, the results can be riveting. One of my recent favorites is "Desperate Measures" by Atul Gawande, which appeared in *The New Yorker* in May 2003. For his subject, he chose what he calls "the murky, violent territory of surgical innovation"—baldly, how many people you're willing to torment and kill in order to test a new theory. Despite its gratifyingly lurid slant, this story would be nowhere without the troubling figure of Francis Moore, a pioneer of organ transplantation who later in life rejected the brutal, anything-for-science approach of his early years. Reading it, I grasped for the first time the appalling human sacrifice that accompanied the halting development of the modern science of organ transplantation (98 of the first 100 heart transplant recipients died within six months) and, by extension, much of late-twentieth-century medicine. We have since become more cautious, Gawande explains. Now a single death can halt all human gene-therapy experiments. But at the end Gawande, a surgeon himself, admits that it's the young Moore he misses, "the one who would do anything to save those who were thought beyond saving." There was a time, he writes, "when we'd have been trying something, anything—and maybe even discovering something new."

6

Finding a Voice and a Style

DAVID EVERETT

David Everett directs the Master of Arts in Writing Program at the Johns Hopkins University, where he teaches nonfiction and science-medical writing. His reporting and writing have won many local, state, and national awards, including honors from the Society of Professional Journalists, the National Press Club, the Overseas Press Club, the Michigan Audubon Society, the Associated Press, and the University of Missouri. In more than two decades as a newsman for the *Detroit Free Press* and Knight-Ridder newspapers, he has reported from 23 states and 11 foreign nations. He has written about the environment, energy, politics, economics, government, and labor, and he has been a copy editor, city editor, Washington correspondent, investigative reporter and editor, and a contributing author of three books. His journalism, essays, humor, and fiction continue to appear in newspapers, magazines, and literary journals and online. David began teaching journalism and writing in 1986, while still a practicing journalist. He lives in the Washington, D.C., area.

I once took a graduate course, from a well-published and finely educated writer, on the topic of voice. In the first moments of the class, several of us audaciously asked the instructor to define the term. A few minutes into her answer, I sensed confusion in the classroom. After 10 more minutes of wandering discussion, it became clear that our teacher couldn't handle this most basic query. She knew it when she read it, she said to our amazement, but who could hope to define voice or its literary twin, style?

Today, after years of teaching voice myself—and of continuing my own writing—I finally understand my instructor's confusion. While all writers crave an individual style, and while we yearn for a distinctive voice for ourselves or the subjects we profile, those goals remain among our greatest challenges, and even experienced practitioners can retreat into debates over their mystery. Many science writers also must contend with journalistic precepts that subjugate or even eliminate individual style. In this chapter I review the complications and examine the tools of voice and style, concluding with exercises that should help writers identify and hone their own.

When writers for the *New York Times* or the Modern Language Association or the *New England Journal of Medicine* talk of style, they often mean the strict rules of spelling, punctuation, abbreviation, and other usage as set forth in hallowed style manuals. Style is also used, more colloquially, to describe writing according to purpose or profession: academic, scientific, journalistic, digital, bureaucratic, literary, postmodern, and so forth. For academics, style has classical roots in Aristotle, Cicero, and that granddaddy of Rhetoric, Hermogenes, who rated style as grand, middle, or plain. Writer Ben Yagoda, in his *The Sound on the Page: Style and Voice in Writing* (2004), defines style as how a writer "uses language to forge or reflect an attitude toward the world."

For the purpose of this chapter, let's define *voice* as *a writer's personality on the page. Style* is the personality imposed on our writing by outside rules and/or our own techniques and mindset. *Voice* is an individual writing personality, whether distinctively our own, one we recount or create, or, sometimes inescapably, both.

In these definitions, *style* is what would differentiate renditions of the American National Anthem by Barbra Streisand, Eminem, Charlie Parker, Buckwheat Zydeco, and Placido Domingo. The tune and words are the same, but what a difference style makes. *Voice* would be when you hear the same patriotic song performed by Parker, Ella Fitzgerald, Louis Armstrong, Billie Holiday, and Wynton Marsalis. It's all jazz—the same *style*—but they're all different *voices.* In science writing, style and voice mean we could assign the same topic to Natalie Angier, Oliver Sacks, John McPhee, Diane Ackerman, Edward O. Wilson, David Quammen, and Annie Dillard, and the results could be as different as hearing that high C "land of the freeeeee" from Eminem and then from Billie Holiday.

For writers, style and voice flow from straightforward elements such as rhythm, punctuation, verb tense, word choice, sentence construction, adjectives and adverbs, and lack of adjectives and adverbs, as well as larger artistic mysteries of attitude, tone, structure, topic, and perception. With voice, what we do not write is as important as what we do. With style, we make readers feel a certain way without mentioning that feeling. Some examples:

Paleontologist Stephen Jay Gould employed an almost Victorian formalism in his writing, to underscore his wondrous view of the smallness of humanity confronted by the might of anthropological time. His nearly contraction-free columns in Natural History magazine liberally included "rather," "moreover," "I will confess," and "it seems," plus the positing of rhetorical questions to which his answers already seemed clear.

Poet and nature writer Diane Ackerman also favors rhetorical queries, but she extends them with strings of phrases and images, adding linguistic flavor and emotion, pushing deeply into each detail, engaging her full poetic instincts, broadening metaphors and empowering emotions, developing each sentence's rhythms, piling on and on, lyrically, beautifully, until we ask, how does she sustain it all? She then concludes (as I'm doing here) with a brief declaration.

Nature essayist David Quammen combines numbers, theories, and chemical names with blunt memoir and adventurous word choice to probe his life, past and present, amid the inherently metaphorical outdoors. He once contrasted his early adulthood to the formulaic measurement of glacial ice flow.

In his articles and essays, food writer Alton Brown uses the same quirky language, crystalline explanations, and self-deprecation that define the on-air style of his television cooking shows. "A big bowl of goo," Brown writes in a *Bon Appetit* article on the science of baker's yeast and his first childhood experience with it. "It was big," he added, "it was sticky, oozing, and kinda smelly." With "kinda" and "way cool" and "plain ole dead" and his trademark first-person style, I could practically hear Brown speaking on the page.

Many writers mistakenly believe that voice and style arise only from beautiful writing. But a trait as simple as clarity, especially when clear thought matches lucid expression, can create its own voice. Consider the "highlighter test." I developed this teaching moment by ruining entire copies of Annie Dillard's *Pilgrim at Tinker Creek* and John McPhee's *The Pine Barrens*. In neon yellow, I highlighted the obviously lyrical passages or phrases in each book. A quick ruffling of pages demonstrates swatches of yellow on most Dillard pages, while the yellow flashes are rare for McPhee.

Does this mean Dillard is a better stylist than McPhee? Only if you focus on prose poetry. A deeper analysis of McPhee shows that his style and voice are defined by brilliant structures and striking clarity. Few readers, including those expert in geology, ecology, engineering, and other fields he explores, would question McPhee's research or metaphors. His explanatory and structural mastery assumes, as he once said, that his readers are smarter than he is, and his voice is as solidly subtle as his viewpoint.

In contrast, we have the famous opening scene in *Pilgrim*, in which a home-coming cat tracks its bloody paws on a sleeping Dillard. As beautifully written and assertively expressed as the scene is, readers have since learned that it fails the accuracy test of a middle school journalism class. Turns out Dillard didn't own a cat at the time, and she did not, therefore, grieve when said phantom cat died, as she later wrote. Her *Pilgrim* Pulitzer was for nonfiction, but we wonder how much is "non" even as we are impressed with her poeticism and enthralled by her relentless philosophizing.

For many science journalists, the quest for objectivity stymies two of the most valuable tools of style and voice—opinion and emotion. In U.S. magazines with a reputation for the best writing (*The New Yorker, Outside, The Atlantic Monthly, Vanity Fair, Harper's*, among others), most contributors rely on a freedom of opinion and expression not permitted in daily newspapers or other magazines. That's why newspapers and news magazines, even those that regularly win Pulitzers, sometimes can't match the best writing in our most literary magazines and narrative books. Emotion sets our voices free.

Remember, too, that the style of some writing is determined as much by the publication as by its writers. Although more individuality is encouraged today than in the past, *Time* and *Newsweek* are still defined by their breezy quips, instant analysis, and inherent readability. Compare that style with that of another weekly, *The New Yorker*, which most often presents pyramidal narrative, assertive intellectualism, and a density of sophisticated language. A good writer can adopt a house style and still strive to be uncommon.

A science writer also knows that style must fit content and purpose. You should use different styles for a news report on the discovery of life evidence on Mars, an investigation of cost overruns on the latest Mars mission, and a philosophical contemplation of a possible human expedition to the Red Planet. It's also fine to change your trademark personal style and voice, as you develop and change as a writer.

Here are a few exercises that should help you, first, to understand more about how style and voice are created and, second, to find or hone your own writing personality:

1. *Study strong voices.* Learn to identify specific technical elements of voice and style. Read the text and make lists; underline words or phrases. It gets easier the more you do it. As an example, Dillard's *Pilgrim* has a fascinating combination of lyrical and clunky language to present her views on the cruel beauty of the struggle for survival in nature. "Eternal chomp," she calls it. I once spent an hour underlining examples of this courageous technique in her book. Other strong voices in science writing include Ackerman, Gould, and Ian Frazier. You also might record and study a speaking voice. For instance, the current president of the United States dis-

plays distinctive vocabulary, diction, and expression. Can you identify the five or six traits that define the presidential voice?

2. *Imitate.* Van Gogh learned to paint by copying Rembrandt and Delacroix; you should try copying too. Write a paragraph or scene in the style of, let's say, Diane Ackerman, with the goal of a reader being unable to tell the difference between your work and hers. Exchange and discuss imitations with writing compatriots. Notice how a small change in word choice can alter a voice; notice, too, how style and voice cannot be divorced from content. One verb will sound more like Gould than like Ackerman. As you experiment with different imitations, pay attention to which ones seem the easiest for you to create, or the most difficult. That comfort level is an important clue for the next set of exercises, when you begin to sharpen your own style and voice.
3. *Freewriting and journaling.* Freewriting involves putting your fingers to pen or keyboard for a certain number of minutes each day—10 or so, to begin—and writing about the first thing that comes to mind and continuing to write about it without lifting your fingers and not worrying about punctuation or speling or language or even how to express yourself but sticking to the subject and trying to relate the first thoughts that come into your mind and then to begin to stop thinking so much about what you're writing and the rules and all that and begin to realize that you're seeing some patterns here and there, maybe some rhythms, maybe some words choices and structural similarities that, after daily freewriting for two weeks, you go back with a beer or cup of coffee and see a trend or something that is distinctly *you.* Another option is a journal, in which you write about one moment from every day of your life for a month and describe, as deeply as you can, how you felt during that moment, without worrying about how you express yourself or whether anyone will read it. The goal for both exercises is to let your instincts emerge and then study the results as closely as you did in exercise (1) above. Underline your own writing; list your own techniques.
4. *Tape yourself.* Have someone record you without your knowing it, in a meeting or during dinner-table conversation. Analyze how you speak; consider whether that is the way you think. If so, it likely will touch on the way you do or should write, without the professional restrictions of style manuals or editorial dictate.
5. *Revise for style.* Pick three paragraphs from the *New York Times* science section and rewrite them in the style of *Family Circle*

magazine, a graduate term paper, *The New Yorker*, a newspaper editorial, a memo for a federal agency, and, finally, an e-mail to your 16-year-old niece. Which was easier? What techniques did you use? Which did you avoid?

6. *Don't write what you know.* Especially at first, try avoiding subjects you are close to. I once had a student whose job was to write about AIDS for a government agency. She chose to write about AIDS in class, too, but found it difficult to break out of the bureaucratic style required at work. So I suggested she write about anything but AIDS. It worked; she began to write with more freedom and flair. Later, if you decide to follow the classic advice and write what you know, you might try doing it from a wildly different perspective to help detect your own style. If you are a news journalist or corporate/agency writer, for instance, compose (privately) an editorial about the persons or topics you cover. Freed from the conventions of your profession or workplace, you should begin to use techniques that reveal genuine voice. Notice how passion and opinion can be essential tools for style. What do you detect in these exercises that is different from your everyday writing?
7. *Examine your personal letters or e-mails.* This writing often is designed to express feelings rather than ideas, to speak casually rather than publicly. You may therefore display hints of your voice more easily than in the serious writing for which you strive professionally. I recently wrote a quick e-mail to a friend about a Cub Scout meeting, of all things, that was more fun to write and, I suspect, to read than the various memos to the dean's office that my academic job sometimes requires.

In the end, your writing personality is confirmed by how comfortable you are with it. Amid all the exercises, text analysis, and rhetorical debate, you will feel more attuned to that personality the more you use it. Even if you can't define either term, the more you give voice and style to your words, the more singularly individual a writer you will become.

Part Two

Choosing Your Market

First of all, congratulations on a brilliant decision.

Since you're reading this, you're probably a general journalist who is switching over to science coverage, a scientist who is switching over to journalism, or a student hoping to become a science journalist. Regardless of which you are, I predict that life will improve. It certainly did for me, after I made the jump from Boston bureau chief of the *New York Times* to science and health reporter for the *Boston Globe.*

As I contemplated crossing that divide, it helped me most to hear: Don't be daunted. You can do it. My idol, Bill Nye The Science Guy, even wrote me an inspirational note that I posted on my cubicle wall, assuring me that despite my ignorance, I understood the PB&J—"passion, beauty, and joy"—of science, and that was what really counted.

Not only *can* you do it, you *should.* In my not-so-humble opinion, science and medicine are producing the most exciting and significant news in the world today. Presidential campaigns, Mideast wars—so much of general news moves in cycles. Science moves in more of a vector forward, and that progress toward knowledge is the closest thing there is to humanity's job description. How's that for being at the center of things?

Enough pep talk. Now shop talk.

In these chapters, some consummate masters of the craft's various media share their wisdom with great generosity and great economy.

Many of their tips are specific to a given medium: in chapter 13 Carl Zimmer lays out what makes a good book proposal, and in chapter 11 Joe Palca talks about what makes a radio story work.

And if you are still deciding exactly how to ply the science journalism trade, they offer some specific hints on that as well. Writing for scientists on the Web is an expanding market, Tammy Powledge reports in chapter 15, while small-newspaper science reporters are rare and ever fighting for survival, according to Ron Seely in chapter 9.

But these disparate descriptions also carry an overarching message, one perhaps best summed up by what a Hollywood friend of mine once told an eggheady screenwriter of dull movies: Dare to be sexy.

Readers must be seduced into reading a science story—just as with any other kind of story, only more so, because they're likely to have to put in more work to understand it. (A favorite editor adds: Dare to be simple, too. You almost can't overdo the simplicity.)

So call it seduction, or call it salesmanship—all stories need it, but various media require various kinds. As do various audiences—whether scientists or lay folk, Web clickers or traditional breakfast newspaper readers.

Colin Norman of *Science* shares some secrets for appealing to scientists: They read enough dry papers; what they like are articles that provide context and color, gathering bits and pieces of information into significant trends and painting portraits of important figures.

As for appealing to laypeople, advice abounds here as well, but I'd boil it down to two main concepts:

First, significance—explain early and often just why this scientific news is worth your audience's attention. (My personal favorite is: You, the reader, will gain new insights into your own nature that will change your self-concept forever after.)

It can be quite a challenge to squeeze the significance into a concise enough form to go as high as it should in the piece. Ron talks about making something called proteomics readable for an audience with a minute and a half to devote to it. But it is an endeavor both noble and necessary.

The second is what you might call juiciness. I like the juice image because it combines two elements: What you write must be appetizing in a human way—the characters must be somehow colorful, the action compelling—and it must be well squeezed, with loads of pulp thrown away. Yes, it hurts to discard work that it took many megahertz of brainpower for you to absorb and understand. But such ruthlessness is well rewarded.

When Mariette DiChristina writes in chapter 16 about what an editor looks for, she mentions nothing about giving a reporter credit for mastering complex science. What she—and any audience—cares about are clarity, good anecdotes, a smooth flow, that kind of thing.

Of course, science journalists in all markets must also practice a kind of anti-salesmanship, what Lee Hotz calls the gatekeeper function. In chapter 8, he writes about a fascinating anthropological find that a journalist of his caliber (my words, not his) could surely write onto the front page of the *Los Angeles*

Times. But was it scientifically valuable enough to merit such play? It was up to him to figure that out.

That gatekeeper function adds some cool water of conscience to all the salesmanship. It is the "Will you respect me in the morning?" part of the seduction: Will you still be credible if you overhype a story?

One other aspect of story-sales that several writers here mention: the package. Alan Boyle writes in chapter 14 that "We're all Web journalists now." At the very least, virtually every science reporter has to think in some visual, video, or audio way.

Not all of us go as far as Alan must at MSNBC, thinking of quizzes, blogs, and photo galleries to go with our stories. But images are so important in science and science reporting that MIT held a special conference on the topic a couple of years ago. Whatever medium you choose, train yourself to think about graphics and other visual aids and you'll add yet another dimension to your work.

To end on a technical note, these pages are full of technical hints, and though they are not exactly trade secrets, they strike me as truly gracious. Lee Hotz has just saved me hours of testing digital recorders, for one thing, and Carl Zimmer may save you days of struggling with footnotes with the software he recommends. Similarly, Kathryn Brown explains in chapter 12 to those of you embarking on a freelance career how to diversify, negotiate, manage your money, and stay sane—all while you're working alone at home in your gym clothes. And Janice Tanne, another successful freelance (who neglects to mention her sartorial habits), in chapter 9 lets you in on the real nuts and bolts of the business, beginning with the need to find out if the editor to whom you're addressing your query letter is a woman or a man.

Joe Palca shares a colleague's tried-and-true recipe for getting comprehensible interviews out of scientists: Tell them to pretend you're a drunk potential funder who won't give them a penny unless they promptly explain what they're doing.

But most gracious of all, these distinguished journalists write with thoughtful candor about what it is like to be them.

And shining through their tips and caveats is a whole lot of the PB&J of science journalism. So it seems not only possible to aspire to emulate an admired byline, but eminently worthwhile—fun, even.

CAREY GOLDBERG

Carey Goldberg is a health and science reporter at the Boston Globe. *She was previously a staff reporter for the* New York Times *(1995–2001) and a Moscow correspondent for the* Los Angeles Times *(1990–1995).*

7

Small Newspapers

RON SEELY

Ron Seely is the science and environment reporter for the *Wisconsin State Journal*, Madison's morning daily and the state's second-largest newspaper. He covers breaking science news from the University of Wisconsin and travels the state and the upper Midwest writing about current environmental issues. He teaches introduction to scientific communication in the UW Department of Life Sciences Communication.

Some days, now that I have crested 50, I find myself surprised to be, of all things, a science reporter on a daily newspaper in a small but sophisticated city and immersed every day in a world of stem cells, radio-collared bald eagles, flakes of Martian meteorites, and strange deer diseases.

I can't imagine a place I'd rather be at this point in my life, though sometimes the haphazard way I got here, the serendipitous nature of it all, makes my head spin—not to mention the fearful task of trying to make something called "proteomics" understandable for an audience that has maybe a minute and a half to read what I've written.

After all, I made a terrible mess of the frog I was supposed to dissect in high school all those years ago. And trying to figure out exactly what Mendel was doing with all of those pea plants drove me nuts. Math? Well, suffice it to say that my problems with algebra and equations are what drove me to a career in journalism.

Still, here I am, settled in Madison, Wisconsin, in the upper Midwest of America, carrying around business cards that identify me as science and environment reporter for the *Wisconsin State Journal.* In a recent, typical week, I researched and wrote a column about robotics, pounded out a quick news feature about why the human body gets cold in winter, made pesky phone calls to state officials about why they aren't enforcing the state's new pollution law, and churned out news briefs on everything from clouds of ionized hydrogen in the solar system to a pollution permit hearing for a local manufacturing plant.

Between assignments, I had time to give a talk about science writing to a local high school biology class and to win one of the ongoing games of Scrabble in the cubicle at the newspaper where I make my workday home.

Not a day goes by that I don't worry about losing this good job. That's because full-time science reporters on small to mid-sized daily newspapers in this country are an endangered species. The *State Journal* is the second-largest daily newspaper in Wisconsin, with a daily circulation of 95,000 and a Sunday circulation of 150,000. This, in fact, is a state full of good daily newspapers in places such as Green Bay and Eau Claire and La Crosse. Yet there are few other reporters who do nothing but write about science for their publications.

All you have to do to understand this trend is think about the writers you ran into at the last meeting of the National Association of Science Writers. Chances are they work for universities in public information offices or for large metropolitan papers or magazines, or they freelance or write books. How many science writers, however, did you meet from daily newspapers in places such as Nashville or Peoria or Fort Wayne or Aberdeen?

Chances are, not many. Which is a shame. At a time when science in one form or another permeates every corner of our lives, one of the most important sources for science news—the daily newspaper—seems to be devoting fewer resources than ever to covering the subject. So here are a few recommendations for the feeding and nurturing of science reporters at small and mid-sized daily newspapers. They are meant to be followed by reporters and editors alike.

■ ■ ■

First, if you are a reporter with science writing tendencies or an editor who wants to see science writing emphasized, make the beat indispensable.

I had covered environmental issues for years at the *State Journal,* sneaking in such stories along with the homicides, floods, and small-town zoning disputes that I covered as a regional general assignment reporter. About 10 or so years ago, when our science reporter left for a job at *Science News,* in Washington, D.C., there were mixed signals about what might happen to the job. I quickly suggested that the new science reporter also be given the job of covering environmental issues, which are generally mostly science anyway. I then lobbied for the job and managed to get it. And the beat, science and environment, has proved to be one of the paper's busiest—which is a fine reason for editors to keep it around. As I had hoped when I proposed combining two beats that might not stand alone at a medium-sized paper, together they provide enough breaking news to justify keeping them. Other papers match science and medical or technology beats, which also makes sense and seems a good way to keep science in the mix.

There are plenty of arguments to be mustered for keeping the science beat on papers such as the *State Journal.* At a time when medium-sized papers are scrapping for readers and competing with television and the Internet, science, presented clearly and succinctly, offers the kind of interesting and useful material that keeps newspaper readers reading.

It is a mistake to underestimate the sophistication of today's newspaper reader, yet it happens too often. In their quest for readers, editors on many papers seem to be opting for flash and design, more headlines, and shorter stories.

But the best way to attract and keep readers, especially the readers of small and medium-sized newspapers, is to offer substantial, interesting stories that make a difference in subscribers' lives, that speak to something they care about. Few beats offer more such stories than the science beat.

A number of recent stories serve as good examples and as proof that the science beat provides important and substantial news that readers care about and that they expect to find covered well in the paper that comes to their door each day.

Two years ago in Wisconsin, chronic wasting disease, a fatal wildlife brain illness, was discovered in the state's wild deer herd. This was no small thing, considering the almost iconic role that deer hunting plays in Wisconsin. The opening day of hunting season each November ranks as an unofficial holiday in the state. More than 200,000 hunters, from all over the country, come to Wisconsin to try and get their deer; over the course of a week, they kill about 700,000.

Even more important, however, is the cultural significance of the event. Generations share the tradition of deer camp. Radio stations play deer hunting songs. Blaze orange becomes a fashion statement—just check out the stands in Lambeau Field during a November Packers game.

Still, at its heart, chronic wasting disease in Wisconsin is a science story. As the paper's science reporter, I made the disease and its impact on the state my story. I became the CWD expert. I developed sources at the University of Wisconsin, including wildlife ecologists and the scientists studying prions, the little-known malformed proteins that cause the disease.

But I also spent time in the field talking to hunters and landowners, broadening our coverage in a way that a science reporter on a large metro might not have to do but that is a necessary part of working on a smaller staff. I carried questions and concerns from the field back to the experts at the university and in the state's Department of Natural Resources. Most uncertainty centered on whether there was a risk that the disease could spread to humans. Unlike with mad cow disease, scientists have so far found no evidence that CWD has resulted in any cases of variant Creutzfeld Jakob disease, the human version caused by eating tainted meat. Writing that story required sound explanations of just how prion diseases work, an understanding of proteins and the role they play in our bodies, and an ability to ferret out the scientific studies that were sound and those that were not.

Beyond becoming the staff expert on prion diseases, I also worked hard to connect our coverage to the average reader—the hunter trying to decide whether to hunt, the hunter's wife worried about having venison in her freezer. By using the science to deepen our coverage in this way, I further justified having a full-time science reporter at the *State Journal.* And our regular and accurate coverage of the issue earned the paper the respect of scientist and hunter alike.

The work on chronic wasting disease later paid off in another way. When mad cow disease was discovered in Washington State, I was well prepared to report the story and its impact on Wisconsin. I wrote once again about the science behind prion diseases and about the differences between mad cow and chronic wasting disease in deer. I worked hard to accurately assess and convey risk—another important job of a science reporter, and information that is meaningful to the general readership of a daily newspaper.

Especially today, so many stories that end up on the front pages of newspapers have to do in one way or another with science. I try hard to keep our editors aware of this and rarely pass up a chance to point out the science angle in the stories we cover. Sometimes I become the lead reporter on the story. Other times I simply contribute a sidebar report about the science.

When stories about flu vaccine shortages started breaking in the fall of 2003, I worked hard to explain to our readers how vaccines work and how decisions are made about what flu strains should be included in each season's vaccine. One of my stories explained the flu tracking system, taking *State Journal* readers from the suffering flu victims who come into small-town Wisconsin clinics to the State Laboratory of Hygiene and from there to the Centers for Disease Control and Prevention in Atlanta. The report not only explained the science of the flu but also connected Wisconsin readers to the national story in a more personal and informative way.

A good portion of my work involves keeping track of the research that comes out of laboratories at the University of Wisconsin in Madison. The university provides not only valuable sources but also remarkable stories about everything from human embryonic stem cells (the University of Wisconsin is a world leader in such research) to brain imaging and the science of emotion. Having a full-time science reporter allows the *State Journal* to pick up more of these fascinating stories, not only the major research projects that are published in the well-known journals but also the less noticed work that might not get covered were the paper relying on general assignment reporters for science coverage.

Most of the time, there is simply too much to write about. But that's a good thing and more evidence that a science beat is worth any newspaper's money and effort. And I like to think that many of the stories I cover would not have been given the attention were I not around to prod the editors. I doubt, for example, that we would have bothered to cover a talk by Harvard's Edward O. Wilson, even though the story turned out to be a timely and interesting discussion of biodiversity in our own backyards. It is unlikely anybody would have bothered to travel to Chicago to listen to a talk by Stephen Hawking, to localize the impact of weak federal mercury regulations on Wisconsin's lakes, or to interview a strange fellow who truly believed he had invented a perpetual motion machine in his basement.

I know that readers appreciate these efforts. One great pleasure of working on a newspaper in a small city is having coffee at a local café and noticing as those around you pore over the paper for which you write. They read and nod and frown and laugh and, sometimes, share a bit of news with the person next to them. I remember, even before I started covering science, sitting in a coffee shop in a small town outside Madison and listening, fascinated, as three old farmers tried to figure out how much horsepower it must take to loft the shuttle into space.

People care about the news in their local papers. I've known this since starting in the business some 30 years ago. For a couple of years, I was the farm reporter for an even smaller daily in the middle of Illinois corn country. I became accustomed to farmers in their muddy coveralls tromping into the newsroom to share news about everything from the weather to the vegetable in their garden that resembled Richard Nixon.

In some sense, it is still that farmer I'm writing for when I write science here at the *State Journal.* This closeness to the people who read your work makes the job both more demanding and more fun. It's nice to get calls in the newsroom from somebody who wants to know what kind of bird has landed on her bird feeder or what that bright star is up there by the moon.

Of course, working on a smaller newspaper has its drawbacks, especially at a time when daily circulation is dropping and most papers are struggling to make money. Budgets always seem tight, so money for travel isn't easy to come by, even for professional conferences or training. But there are a number of training programs and fellowships out there that will cover part or all of your expenses. The website Journalismjobs.com will provide you with a good list of such opportunities.

Slim budgets and smaller staffs also mean that your time isn't always spent doing just science reporting. The science reporter gets no exemption at the *State Journal*, for example, from having to pitch in and help compile the annual business tab or cover a local county board race or work the Saturday general assignment beat once a month.

Still, such nuts-and-bolts journalism keeps one humble, and though I am a science reporter, I am first a working newspaper stiff and as addicted as anyone on the staff to breaking news and the big stories and the feel of a paper in my hands as I flip through the pages.

Science has a place in such a time-honored medium, and I like to feel that, in Madison at least, the folks at the coffee shop are getting their money's worth of science news when they read their morning paper.

8

Large Newspapers

ROBERT LEE HOTZ

Robert Lee Hotz covers science and technology for the *Los Angeles Times*. He has been a science writer for most of his newspaper life, which began in 1976 at a small country daily in Virginia's Shenandoah Valley. He has twice received national reporting awards from the Society of Professional Journalists and three times won the science writing award given by the American Association for the Advancement of Science. He was a Pulitzer Prize finalist in 1986 for his coverage of genetic engineering issues and again in 2004 for his stories about the space shuttle *Columbia* accident. And he shared in a staff Pulitzer in 1995 for the *L.A. Times'* coverage of the Northridge earthquake. Lee is an honorary life member of Sigma Xi, the scientific research society, and vice president and president-elect of the National Association of Science Writers.

It was a nice rock, as rocks go—a substantial chip of rose-colored quartz gleaming with flecks of crystal—but not the sort of stone that might grace a starlet's ring finger.

Even so, curators at the American Museum of Natural History in New York had given it the kind of showroom treatment Tiffany's might lavish on its rarest diamond solitaire: a special exhibit case, dramatic spot lighting, and even a name designed to stir the imaginations of onlookers.

The rock was a 350,000-year-old hand ax. The Spanish archaeologists who discovered it called it Excalibur. And they claimed it was the earliest known evidence of the dawn of the modern human mind.

Found among the skeletal remains of 27 primitive men, women, and children, the ax might be the earliest known funeral offering, its discoverers contended. If so, it was 250,000 years older than any other evidence that such early human species honored their dead.

As a reporter, I was in a bind.

Discovery of the rock offered an opportunity—the potential news hook—for a fascinating story. But it posed a series of thorny questions that I had to resolve before I could, in good conscience, publish a story about the find. They are the questions that arise with every newsworthy scientific development. They center on the validity of the work, its importance to the general public, and whether independent scientists can vouch for it.

There also are practical considerations. How much of a reporter's time is it worth? How quickly can the story be turned around? Is there enough material for a graphic? Can we get a photograph? How much space does it deserve? Does it have a chance of getting on page one?

The claim being made by the Spanish archaeologists was certainly provocative and, no doubt, sincere. But how reliable was it?

The study of human origins is a field defined by the paucity of evidence and conflicting scientific claims. As one distinguished paleo-anthropologist told me wryly, "The dividing line between reality and paleo-fantasy is very narrow."

Acting as a gatekeeper to sort the sense from scientific nonsense, a science writer ordinarily can spend almost as much time chasing down a misleading claim as publicizing valid work. In this instance, I had to ask myself whether there was anything besides the speculative enthusiasm of the archaeologists who made the discovery to support such an extraordinary claim.

By itself, the rock offered nothing that directly revealed its significance. Like so many prehuman remains, its importance was all a matter of context: the circumstances in which it was found, the age of the deposits around it, the interpretation placed on the find by its discoverers, and, in this instance, the implied endorsement of the American Museum of Natural History, one of the nation's oldest and most distinguished research enterprises.

To pursue the story, I interviewed the archaeologists themselves, going back twice with follow-up questions. They were articulate, charming, dedicated, authoritative, and infused with the romance of possibilities.

Some of the most respected authorities on human origins are members of the museum's research staff. I also interviewed them at length. Although they were emphatic about the overall importance of the finds on display, they were more reserved about the significance of the ax itself and tended to steer the conversation delicately away from the subject.

As part of the exhibition, the museum curators had convened a two-day conference about the state of research into European prehistory. Journalists were not allowed to listen in on the actual scholarly debates. Instead, a special session was convened for reporters. Several scientists summed up the proceedings. They also distributed a carefully worded consensus statement about the importance of the specimens on exhibit.

Such reticence began to trigger mental alarm bells.

Often, a noteworthy fossil find is trumpeted by one of the important peer-reviewed journals, such as *Science* or *Nature*. There was no such prominent publication about this ax in the offing, I learned. But the work would be detailed at a later date in a more minor but respectable academic journal in Europe that specialized in human anthropology, I was told.

In effect, the museum exhibit itself was the public announcement of the claim. The exhibition's scholarly catalogue documented the finds in 147 pages of authoritative, bilingual text. The ax, however, took up just two pages of the tome, most of which were devoted to a large color photograph. There was a single terse paragraph of supporting text.

Clearly, there was no suggestion of fraud or conscious deception. I did wonder, however, whether the Spanish archaeologists had in good faith simply overreached. Other researchers, too polite perhaps to naysay their claim in public, were damning it with faint praise.

By this point, I had invested two days in researching the story, and under other circumstances, I might have given up on it entirely as being too speculative or offered a brief about the exhibit to our travel section. One advantage of working at a very large newspaper like the *L.A. Times*, however, is that a reporter often has more freedom to scratch his curiosity itch and pursue a tantalizing lead with the implicit understanding that breaking news may at any moment interrupt his work and plunge him into deadline coverage of unfolding news events.

At my first newspaper—a country daily with a circulation of 11,000 and an editorial staff of 10—I was expected to write three or four stories a day, then take obituaries over the phone or type up wedding announcements. At the *L.A. Times*, a reporter sometimes can spend weeks or even months on a single story, able to pursue it with a tenacity and depth beyond what a smaller paper can usually afford. In particular, the science writers at the *L.A. Times* have considerable latitude in choosing their assignments.

The *Times* employs about a thousand reporters and editors, more than two dozen of whom specialize in coverage of science, technology, medicine, or the environment. The science editor reports directly to the managing editor and attends the daily meetings where story play and space allocation decisions are made.

Every story has to fight to find its place in the paper. Competition for a place on the front page is especially intense. Normally only seven or eight stories can be displayed on the front page of the *L.A. Times*.

Science writers often argue that the news of science must be sheltered in a special weekly section, in the belief that it cannot easily compete for space with the hard news of the day. At the *L.A. Times*, however, we are confident that the important news of research can easily hold its own in the rough and tumble of a daily news report. Even so, the challenge at any large metropolitan daily is to find the science story that can break into print, either by reason of its practical impact to readers or its value in the growing store of human knowledge, or by its ability to pull readers into the curious wonder of the research realm.

I was reluctant to let go of this story just yet.

Talking it over with *L.A. Times* science editor Ashley Dunn, I began to realize that the uncertainty itself might be the core of the story. It was a curiously ambiguous claim. The researchers had taken the prehistory of the mind right to the edge of what could be deduced from an inanimate object.

Now I had reached a point in my reporting when I had to draw more directly on my own resources. Any reporter has to be a packrat when it comes to collecting information. Some of us raise the practice to the level of a personality disorder. Like most of my colleagues, I routinely keep scores of files on topics that might one day blossom into print.

In that regard, the advent of the electronic newsroom has been a blessing. As a journalist, I am a card-carrying member of the digerati. Every aspect of my work is informed and organized by computer tools. There may be no information storage and retrieval technology quite as robust as pen and paper—I always carry a reporter's notebook with all-weather writing paper so my notes won't smear in the rain—but electronic tools can do much to ease a science writer's job.

Any computer program has a steep learning curve, so the most important thing is to find a program that suits your needs and then stick to it. Work habits should be just that—habits—a collection of ingrained techniques that come as easily and automatically as touch typing.

I keep track of hundreds of sources in a contact management computer program, not too different from the computer programs that salespeople use to keep track of customers. Many people use Microsoft Outlook for this purpose, mostly because so many computers come with it already installed, but for the past 15 years I have relied on an information management program called Commence produced by the Commence Corporation. It allows me to cross-reference people by story or expertise and to link them to ongoing projects, to-do lists, or calendar appointments. It allows me to customize these databases on the fly to suit a particular need and keep them all in sync across the different computers I use at home, in the office, and on the road. It also updates the contact files in my Palm Pilot.

For interviews, I use an Olympus digital voice recorder. The audio quality is better than conventional tape. There are no moving spools to jam, no tapes to lose or mistakenly erase. The removable memory card can hold up to 22 hours of interviews. To improve the quality of those recordings, I also use an external stereo microphone about the size of a bottle cap.

I file all my recorded interviews on my office computer, where I can easily archive them for retrieval and transcription. Since the interviews are recorded in a digital format, I link the recorder to the computer with a cable and download the audio files with a simple click of the mouse like ripping a music file. I can then listen to the audio files, edit them electronically, index them, or copy them onto a CD-ROM for sturdy archival storage.

A digital recording is also faster to work with than a conventional tape that must be mechanically wound and rewound, during the effort to cull accurate quotes. The sound quality is usually superior, too, because there is no tape hiss or machine noise during playback. There is no tape to tangle or snarl, no magnetic oxide to flake away with time. I have noticed that my radio colleagues favor Sony minidisc recorders, which record on tiny CD platters, or digital audio tape (DAT) machines, both of which combine broadcast quality audio with the ease of electronic editing.

In the hope I could end the drudgery of laboriously transcribing an interview, I have experimented with several voice recognition programs that can transcribe voice files directly to text, such as Dragon NaturallySpeaking, produced by ScanSoft. So far, however, these programs can be trained to master only the speech patterns of one voice. They can't begin to handle the vocal variety of all the people a science writer is recording. I still have hopes, though.

To round out my digital toolkit, I recently started carrying a digital Canon camera in my briefcase. I take pictures during lab visits or interviews as a form of note-taking, to capture details of places and people that I might not be observant enough to jot down.

Although many newspapers are still struggling to come to terms with the multimedia potential of the online universe, I like to be able to offer up snippets from my digital audio files and my electronic photographs as supplementary material to round out a science news story posted on a website or in a blog.

To collate news clippings, research papers, Web pages, and anything else that might one day become indispensable on deadline, I use a free-form database called askSam, produced by askSam Systems. With it, I can type information directly into the program or import it from almost any electronic source, including Web pages, PDF files, e-mail, text files, spreadsheets, or other data base programs. The program is designed to automatically turn information into a database that can be easily searched in more ways than any other database program I have tested. I use it to build electronic archives on topics of general interest to me, such as neuroscience or the U.S. space program. I also use it to collate source material for major stories.

I keep copies of such work files on portable "thumb drives" (also called USB flash drives, or jump drives)—one on my keychain and another in my briefcase—so that when I am traveling I can plug one into any computer I might have to borrow and have timely access to the information I might need.

I turned to those resources now, to recall those researchers who were expert in related fields, who might be able to put the discovery of the ax into a broader context about the nature of scientific inquiry. Soon I was conducting follow-up interviews with independent experts in Spain, at the Smithsonian Institution in Washington, D.C., at Stanford University, at the Museum of Natural History in Paris, at University College London, and at the Center for Human Evolutionary Studies in Cambridge, England. Playing telephone tag across so many time zones can be expensive and exasperating. I often use e-mail as a way of introducing myself and arranging interviews.

Over the course of a week and a half, a story about the archaeology of the mind started to take shape. The rock neatly illustrated the problem of documenting mental evolution. The interviews with independent researchers in the field had given me enough confidence in the find to bring it to my readers' attention.

At the same time, I began working with our art department. For a science news story, informational graphics are an essential element. The *L.A. Times* places a premium on page design, and the story that offers arresting elements for a designer to use in laying out a page can benefit.

To help put the ax in context, we created a timeline that noted milestones in the history of the evolving mind. We also prepared a map showing the site of the excavations. From the museum staff, I obtained color slides of the ax and of researchers working in the eerie cavern in which it was discovered.

In the meantime, I sent my editor a summary of the story that I intended to file. In this form, the story joins the jostle for space and display. Typically, that summary includes an estimate of the story's length, its estimated time of arrival on the news desk, and any visual elements that will accompany it. The summary also offers a sample of the story's first paragraphs. That way those senior editors who decide how stories will be played in the newspaper can get a direct sense of the story's flavor and style. The quality of the writing can help lift a news feature from the interior of the paper and into the spotlight of the front page.

Next I started drafting. After a day or three, and five or six rewrites, I had completed a 2,000-word draft, ready to be edited. My editor peppered me with thoughtful questions designed to strengthen the story, sharpen its main points, and improve its narrative flow. He also had good news: The ax story was a candidate for the Sunday front page, the day of our highest circulation and, therefore, considered the paper's showcase.

The article began this way:

> NEW YORK To the primitive hands that deftly shaped it from rose-colored quartz 350,000 years ago, a glittering stone ax may have been as dazzling as any ceremonial saber.
>
> It was found in the depths of a Spanish cavern among the skeletal remains of 27 primitive men, women and children—pristine, solitary and placed like a lasting tribute to the deceased whose bones embraced it.
>
> For the archaeologists who unearthed this prehistoric blade, the unique burial site is a compelling but controversial glimpse of arguably the earliest evidence of humanity's dawning spiritual life.

By custom, I arrange to be in the office on the Saturday before a major story of mine runs in the Sunday paper, so that I can be on hand for any last-minute questions or copy desk queries. By the time I reached my office early in the morning on Saturday, February 1, 2003, however, my editor was already calling.

The Sunday front page had been wiped clean by events.

On its final approach to landing in Florida, the space shuttle *Columbia* had disintegrated, spreading wreckage across seven states and killing the seven men and women aboard. We all were mobilized to cover that breaking news story.

It was weeks before we thought again about anything as esoteric as a prehistoric ax.

Due to the change in emphasis, the story, fortunately, was sufficiently timeless. The museum exhibit that had prompted the press conference now merited just a passing mention.

On Saturday, February 22, the story of the quartz ax was published on the front of the *L.A. Times* World section, giving more than three-quarters of the page to the ancient mysteries of the evolving mind.

9

Popular Magazines

JANICE HOPKINS TANNE

Janice Hopkins Tanne has been a successful freelance writer for popular magazines for 20 years. She originated "The Best Doctors in New York" for *New York* magazine and "The Best Doctors in America" for *American Health*. Her articles have appeared in *Columbia Journalism Review, Family Circle, Self, Woman's Day, Child, Parade, Vogue, The Carnegie Reporter, Reader's Digest,* and many other newsstand magazines. She is co-author, with Dr. Lee Reichman, of *Timebomb: The Global Epidemic of Multi-Drug Resistant Tuberculosis* (2002), and she reports medical news from America every week for *BMJ (British Medical Journal)*. Janice has won 11 awards for her medical stories. She is a past president of the American Society of Journalists and Authors, an officer of the Newswomen's Club of New York, and a member of NASW and the Authors Guild.

I like to know how things work—why plaque piles up in arteries, how microbiologists identify different strains of bacteria, how surgeons separate conjoined twins, why some medical centers are better than others. I want to give people information that will help them make better medical decisions. The most exciting way to find these things out, and to inform the public, is by writing for popular magazines.

Most magazine articles begin with a proposal, also called a query letter. For a piece I suggested to *Child* magazine, here's how I opened my proposal:

> Eva Marie is an energetic 4-year-old beauty with big dark eyes, a shining pageboy haircut, and a hole in her heart.

> Yes, doctors said, the hole can be repaired, but the surgery will split open Eva's breastbone and leave a scar down the center of her small body. Eva Marie would spend three days in the pediatric intensive care unit and at least seven days in the hospital.
>
> How can you explain to a 4-year-old what major heart surgery means? Eva Marie's parents hoped to find a way to save her the pain, the scarring, and the lengthy recovery time.
>
> They were fortunate. Surgeons at New York University Medical Center who pioneered minimal-access heart surgery for adults are now using the technique to treat children's heart defects like the one Eva Marie was born with.

The editor liked the proposal and assigned me the story, which ran as the article "Gentle Repair for Tiny Hearts" in the August 2000 issue of *Child.*

The Marketplace

The major magazine markets for health and medical stories are general interest magazines, women's magazines, parenting magazines, health magazines, and science magazines.

With a circulation of nearly 36 million and a readership of 80 million—nearly one-third of the nation—the granddaddy of general interest magazines is *Parade.* It is the Sunday magazine of 350 newspapers. *Parade* has a commitment to health coverage; it runs one major health story a month, plus a health column. It also features special issues on "Live Longer, Better, Wiser," men's health, women's health, and issues keyed to important "disease weeks." Other general interest magazines that cover health issues are *Reader's Digest, The Atlantic Monthly, Harper's,* and many regional magazines, like *New York.*

Women's magazines vary among themselves, but all consider their readers to be busy people who don't have a lot of time to read medical stories. As former *Woman's Day* health editor Jillian Rowley puts it, the typical reader "wants to simplify, do things quickly and easily, but also keep herself and her family fit and healthy. The information we give our readers has to be useful." *Woman's Day* and *Family Circle* are aimed at women in their 30s to early 40s; *More* is for women over 40. And *Self,* according to senior editor Elizabeth Anne Shaw, is for younger women: late 20s to early 40s, who are "really healthy, active, highly educated."

Parenting magazines tend to focus on healthy development and common problems (won't eat his veggies), not on rare conditions, though sometimes they go for the drama, as my story for *Child* demonstrated.

Health magazines include *Health, Prevention, Men's Health, Fitness,* and *Shape. Health* has some popular-psychology stories. *Prevention* often looks at "natural" treatments. *Men's Health* is guy-talk cute, with lots of tips and sidebars. *Fitness* and *Shape* lean toward exercise stories. There are also health newsletters published by medical schools and major medical centers; many of them use freelance writers.

Among the science magazines, the most famous is *Scientific American,* whose readers are mostly men in their 40s with high incomes. News editor Philip Yam says the magazine is serious, but lighter and livelier than it used to be. It covers all areas of science, not just medicine. Like many editors, he says, "I want to be surprised, to see a fresh angle on an old idea." Also in this category are *Popular Science, Discover,* and *Wired.*

Whatever magazine you're targeting, read several issues before you submit a proposal. The magazine's website may have guidelines for writers, or you may get them with a call to the magazine's editorial department.

The Proposal

Your proposal must convince an editor that you have a good story idea, that you write well, and that you're the ideal person to do the story. Most of my queries open with a sentence that becomes the lead of my story. Editors love a query that shows you have found people and experts to interview. In my Eva Marie query, the opening made clear that I'd already talked to the surgeons and had found a patient.

Address the proposal to the right editor. Don't send your query blindly to "Health Editor" or "Articles Editor." Get that person's name from the masthead, and make sure you spell it correctly. With a unisex name like "Brett" or "Leslie," find out if the editor is a man or a woman. The managing editor of a major weekly publication has a unisex name and discards queries addressed to "Ms." because he says that a writer who hasn't done enough research to discover he's a guy is not somebody he wants writing for him.

Only a few years ago, writers put a neatly typed query letter and some of their best clips into an envelope and mailed it. These days many editors prefer e-mail. But it varies. Some magazines prefer snail mail because they receive so many e-mails and so much spam that many worthy e-mails don't get through. Some editors like snail mail better because they think a proposal and clips on paper are easier to review and pass around.

Find out what editors prefer by calling the magazine and asking. If you e-mail the query, indicate on the subject line that it's a proposal.

After you send it, your query will probably fall into a black hole. Most editors do not respond unless they are interested. Some queries get lost. If you haven't heard in two or three weeks, e-mail or call the editor. Remind her politely about your query and paste a copy of the query into the e-mail. If there's no response after two follow-ups, consider it quits with that publication.

Multiple queries make sense when you have a time-sensitive idea and you're suggesting the story to editors you haven't worked with before. If you do know the editor, you might want to use wording suggested by NASW member Robin Mejia: "I'm offering this to you exclusively. Because of the timely nature of the story I'd appreciate a response by XXX. If I don't hear from you I'll assume I'm free to pitch it elsewhere."

Writers worry that if they do multiple queries, two editors may want it. That seldom happens. If it does, you can take the best offer or do different stories on the same subject.

The Contract

Don't begin work on your article until you have a contract, even when the editor says she's desperate for a story by Monday morning. A colleague started work on a timely story before he got the contract and then—whoops!—the contract said the magazine wouldn't pay until he submitted tapes of his interviews. But nobody had told him he was required to tape interviews, and by then he had already conducted several. Dig in your heels and say you won't start work until you get a contract. That's why there are fax machines.

Magazines used to buy "first North American serial rights," which allowed the publication to publish your story one time only, in North America. Travel writers often resold articles to newspaper travel sections across the nation. Authors of "evergreen" articles suitable for many magazines (like "10 Tips on Keeping Your Living Room Tidy" or "The Most Important Information on Your Resume") often got a nice income from resales.

Things changed with electronic databases and the Internet. Publishers wanted all rights to stories—not so much because they could resell stories to the same markets that writers used to sell to, but because they gained handsome fees from selling the entire contents of their publications to databases.

Warranty and indemnification clauses are another problem. Some contracts require that you pay for the publisher's Wall Street lawyers if someone decides to sue the magazine. They also might ask you to warrant "that the work is not libelous, obscene, or otherwise in contravention of law and does not violate the proprietary right, right of privacy, or other right of any third party." You can't, because those decisions are made by courts. Writers should insert "to the best of my knowledge" in that section. The publisher's lawyers should check for libel if the story is sensitive.

A contract is negotiable. Many magazines have at least two, sometimes three, contracts, and they try first to get you to sign the one that keeps most reprint rights for themselves. If you get one you don't like, try saying something pleasant, like, "My lawyer doesn't like me to sign those all-rights contracts. Could you send me the other one?"

Writers can ask magazines to modify their contract, and many magazines will. Writers' organizations, especially the American Society of Journalists and Authors, provide useful contract advice. Sometimes all you can do with an especially dreadful contract, if the publisher is unwilling to negotiate, is to turn down the assignment and refuse to write for that magazine. I have, and so have many of my colleagues.

Writing and Submitting the Story

Writing an article is a process of gathering too much information and then winnowing it down. I try whenever possible to do my interviews in person, to watch the surgery, to go to the lab. I tape when I can, but I usually listen only to crucial quotes. I take careful notes as well. Tape recorders fail, and they're not as helpful in operating rooms as taking notes of what you're seeing.

As for organizing the story, I started out in newspapers, where you don't have the luxury of time. I learned to outline the story in my head before I sat down to write. Sometimes, however, I'll run into a real bastard of a story that seems to lack a natural organization. Then I try writing the individual parts of the story with the aim of linking them together. Often a structure appears as I do this.

One way of organizing is starting *in media res* (in the middle of things, as the Romans said). You begin at the crisis point and then you go on to explain to the reader just what led up to the crisis, what happened next, and how it was resolved. Another popular method is to begin with a touching story to engage the reader. I call this the "Mildred, a 34-year-old mother-of-two lead." Then you explain how common Mildred's problem is, what the warning signs are, and how it is diagnosed and treated, and include some tips on finding expert help. A variant of this opening is to mention a current news event or a historical discovery, then explain why it happened and what it led to.

Your editor may want revisions or additional information. At some hellish magazines, one editor will make comments and ask for revisions and after you've done them, then the new version is passed on to another editor, who asks you for more and different revisions, and then. . . . At other magazines your story will be circulated to all editors and their comments and requests for revisions will be consolidated before they get to you.

You should be paid upon acceptance, not upon publication. The dreaded *kill fee* should come into play only when all people involved have made their best effort and the article just isn't working. But sometimes there's a new editor who changes her mind and just doesn't want your story. That is when you should get your full fee, since you've fulfilled your assignment but the magazine has killed the piece for its own reasons. Kill fees range from 10 percent to 100 percent; most are 25 to 30 percent.

When your article has been accepted and edited, it will probably be fact-checked by a bright young person who may save you from errors. Give this person copies of your published sources, marked to show what supports what you wrote, and a list of the people you interviewed, with their phone numbers.

Some magazines ask for your notes and also for tapes of your interviews and for transcriptions. You need to know this up front. I don't always tape and almost never transcribe, but if a magazine wants that, I try to get it to pay for transcription just as it pays for travel and telephone expenses. I don't like giving a magazine information that doesn't relate to the story it's paying for—maybe the interviewee told me about something unrelated that would be a story for another magazine.

You should see galleys or some final edited version of your article before publication in case a mistake has crept in during editing that might give readers incorrect and dangerous information or just make the magazine, and you, look really dumb.

The Best Part

The best reward for all this hard work—besides just seeing your name in print and your article beautifully displayed—is getting a handwritten note from someone whose life you have saved. I've been blessed with a few such letters. The most touching was from a California woman who was successfully treated for a dangerous brain tumor thanks to my article in *Parade,* and then went on to have two wonderful children. Stories like this make it all worthwhile.

10

Trade and Science Journals

COLIN NORMAN

Colin Norman is the news editor at *Science* magazine, where he manages, to the extent possible, a far-flung, award-winning team of staff and freelance science writers. He has a degree in Liberal Studies in Science from Manchester University in the United Kingdom, but no formal training in science journalism. He learned on the job—and continues to do so—first at *Nature* as a staff writer in London and Washington, D.C., followed by a stint as a senior researcher at the Worldwatch Institute in Washington, where he thought deep thoughts and wrote about science, technology, and society. After temporarily curing himself of the urge to write books with *The God That Limps* (1981), about the social impact of new technology, he returned to weekly journalism, joining *Science* in 1981 as a writer and editor in the news department.

I once asked a friend who was a political reporter for an influential British daily whether he had a particular type of reader in mind when he sat down at his typewriter (yes, I said typewriter; it was a long time ago). His response: "Somebody who moves his lips when he reads."

At the time, equipped with a mere bachelor's degree, I was starting out as a reporter for *Nature*, a journal read by researchers at the forefront of their disciplines—Nobel laureates, even. My friend's flip remark carried a useful message, which is why it has stuck with me over the years: Don't be intimidated by your readers.

Writing for a scientific journal can certainly be intimidating. A fraction of your readers will know a good deal more about the topic than you do, and a larger fraction will be quick to jump on any mistakes. Yet if you are writing for a multidisciplinary journal like *Science* or *Nature*, and are hoping to entice an astrophysicist, say, to read an article on genetics, you'll need to explain some basic terms—and you'll need to do it without talking down to the scientists who are already the experts. You are also writing for a very busy audience, so there's a premium on good writing. Scientists have a hard enough time understanding the technical papers in the back of the journal, and they will turn the page rather than struggle through a news story if it's needlessly dense. And, perhaps most important, you are setting the context of whatever research you're describing. Your readers can get the findings just by scanning the literature, but what they can't get is how those findings fit into a hot new trend or the way that intense competition drove the research. That's where you come in.

So what makes a good story for a professional magazine? Remember who your readers are: a community of scientists—a relatively specific community if you are writing for a magazine like *Chemical and Engineering News*, or a very broad one if you are writing for *Science* or *Nature*. Like members of any community, they share common interests and concerns—not just about the latest scientific findings, but about the forces shaping the community from inside and outside, including trends, conflicts, personalities, competition, government policies, and, of course, money. A good story for a professional magazine is one that plays to that community's particular interests.

Take bread-and-butter stories about research findings. When *Science* pioneered the idea of journalists writing about research in a scientific journal almost 35 years ago, the very notion was heretical. Only scientists can write accurately and with authority about science, the argument went. We still hear that refrain occasionally. But the news sections of *Science* quickly became the most widely read parts of the journal. The reasons: timeliness, objectivity, context, and clarity. News articles about research will appeal to scientists if they make new results understandable and draw connections that might not be obvious from the dry prose of a research paper. They should also offer other investigators' views on the strengths, weaknesses, and implications of a provocative finding—insights that can help a busy reader filter the scientific literature.

The criteria for what makes a scientific development a potential story for a professional magazine are not that different from those that apply to stories for a more general audience: The findings should be important, preferably provocative, have implications that go beyond a narrow area of research, and so forth. But there are some wrinkles. Certain findings are likely to be of special interest to readers of professional journals: those that challenge the conventional wisdom in a particular field, or that fit into a fast-moving area of research, or that have implications across scientific disciplines. An example that fits all three of those categories is the spate of findings on small RNA molecules in the past few years that have begun to transform our understanding of how gene expression is modulated in many environments. News articles and perspectives about small RNA molecules began appearing in the pages of scientific journals around 2000, and we made this area of research *Science*'s Breakthrough of the Year in 2002. As I write this, the topic has yet to be widely covered in the general media, in part because most of the immediate implications are largely confined to the laboratory. But in the scientific community this is a hot story.

The best places to get such stories are scientific meetings, which offer not only the first public presentations of the latest research findings but also public and private critiques of those results by other scientists. The combination can pay big dividends. My *Science* colleague Dick Kerr, for example, heard a series of presentations at a meeting in early 2003 that offered a startling explanation for several puzzling features on Mars: Periodic shifts in the planet's angle of rotation may have caused water at the poles to be deposited at lower latitudes as dirty snow. In other words, Mars may have experienced a series of ice ages. Dick gathered views on this idea at the meeting, ran them by his many sources, and wrote an article for *Science* in April 2003 in which he coined the term "iceball Mars." A paper describing the evidence was published several months later in *Nature*, and soon "iceball Mars" became part of the planetary science lexicon.

Once you have an idea for a story about research findings that will appeal to the readers of a professional journal and you have convinced your editor that it is worth pursuing, how do you go about writing it? It's tempting to give authority to your writing by trying to sound like a scientist. That's a big mistake. One point comes up repeatedly in focus group meetings we hold at *Science* and in everyday conversations with scientists: Scientific papers are largely impenetrable to scientists in another discipline, and they are hard enough to understand even for scientists in the same discipline. Scientists look to the front of the magazine—the news and Perspectives sections—to provide understanding and clarity. So, avoid jargon, keep acronyms to a minimum, and go light on mind-numbing experimental details. In fact, when you are writing for a broad scientific audience, the techniques are not very different from those you would use in writing for the general public. A reader who needs a greater level of detail will go to the paper to get it.

But you do face a few problems that are peculiar to writing about science for a scientific audience. The most obvious is that some of your readers will be experts and you'll lose them if you pitch the story too generally, yet the bulk of them come from other disciplines and may need some explanatory background. There are a variety of tricks you can use to walk that tightrope. Slip in definitions as asides rather than labored explanations, as if you are simply reminding the reader: "Two teams of researchers announced that they had created a type of matter known as a Bose–Einstein condensate—a cluster of particles that acts like a single, enormous quantum-mechanical object." You don't need to define DNA, but if in doubt, provide a brief reminder. Again, it helps to have a particular type of reader in mind when you sit down at your keyboard: Write about physics for a biologist.

Another difference is that your readers will hold you to a very high standard of accuracy—higher even than when you write for a newspaper. Don't be afraid to keep asking questions until you have it straight, and don't be afraid to keep going back to your sources to clarify points, check facts, and get responses to new information that comes up in your reporting. Mistakes will be pounced on. Because of a mix-up in a caption, we once referred to a crab as a mollusk, prompting a flood of letters along the lines of "if even *Science* can't get it right, it's no wonder the standard of scientific knowledge in the country is so abysmal."

If accuracy is paramount, should you ask a source to read a draft of your story? That's a question that can generate some strong opinions. At *Science*, reporters check facts with sources and sometimes ask them to read a draft, with the express understanding that the draft is confidential and that we are asking only for a factual check. We generally find such reads helpful.

You should also pay attention to crediting key contributions from other groups. One of the more common complaints we get at *Science* is that we didn't mention a paper—usually from a competitor—that led to the findings we are discussing. Clearly, you can't trace the entire intellectual history of a research development, but in the community that reads scientific journals, credit is critical. And if there is intense competition behind a new result, that should be a part of the story.

Indeed, some of the most memorable stories in professional journals are those that focus on intellectual disputes, the competition that drives an area of research, and the personalities involved. This kind of story can be difficult to write, in part because scientists themselves may be reluctant to discuss motivations that conflict with the myth that science is driven only by data and the search for truth. But science is a highly competitive enterprise, filled with lively and interesting characters. Writing about science by focusing only on data would be like covering Congress by focusing only on legislation.

Conflicts can lead you to hot areas of research—the most intense fights tend to be about important scientific issues—and they can also be a way to write about difficult areas of science in a lively way. A classic example of this genre is a series of articles (later turned into a 1981 book, *The Nobel Duel*) written for *Science* many years ago by Nicholas Wade, now a reporter at the *New York Times*. He described the 22-year race between Roger Guillemin and

Andrew Schally to isolate brain hormones, work that involved laboriously grinding up tons of animal brains and tedious efforts to detect minute levels of the elusive peptides. The main reason both scientists persevered, as Wade vividly portrayed it, was mutual dislike and the fear that the other would get there first. The race ended in a virtual dead heat, and they shared the Nobel Prize. Describing this groundbreaking work by focusing on the forces, setbacks, and triumphs behind the data put a human face on the science, one that many scientists could relate to.

More recently, Jennifer Couzin tackled one of the hottest areas of longevity research in the news pages of *Science* by describing the bitter rivalry between Leonard Guarente, a prominent researcher in aging, and his former postdoc, David Sinclair, who now holds different views from Guarente's. The research, conducted in yeast, involves pathways that might explain why cutting calorie intake appears to lengthen life span. The research may have implications for mammals, which also tend to live longer on near-starvation diets, but the biology is so intricate that simply describing it would quickly lose the average reader. Setting it in the context of the mentor–student clash, however, brought the work to life in a way that resonated with scientists across the board.

Articles like these are widely read because they depict forces that shape science—they play to the community's special interests. (Some scientists might demur. A reader once complained to me that *Science* has too much gossip, but he then proceeded to rattle off several recent examples that he had clearly read.) Another popular type of article describes external forces that influence the way science is performed—the politics of agencies that fund research, the growing commercial stake in academic research, regulations, career prospects, scientific misconduct, and public attitudes toward research. Taken together, these external forces are essential ingredients in the coverage of science for a scientific readership.

As you cover science for a scientific journal, think of it in the same way that the *Wall Street Journal* covers business. Just as there's a lot more to business reporting than writing about company news and stock prices, there's more to science writing than reporting about data. Scientists—your readers—have a consuming interest in the scientific community at large. Keep them in mind as you sit at your keyboard.

11

Broadcast Science Journalism

JOE PALCA

Joe Palca is a senior science correspondent for National Public Radio. He comes to journalism from a science background, having received a Ph.D. in psychology from the University of California at Santa Cruz, where he worked on human sleep physiology. Joe was president of the National Association of Science Writers in 1999 and 2000. He has won numerous journalism awards, including the National Academies Communications Award and the NASW Science-in-Society Award. He lives in Washington, D.C., with his wife and two sons.

When I first made the jump from print to broadcast, people kept asking me if I missed writing. The question was funny, but also vexing. I hadn't stopped writing, I was just writing in a different way. And not really all that different, just shorter. But after a time my vexation went away, and I decided the question was a form of flattery. Good radio stories are intimate and personal, where the listener gets a sense of being talked with, not talked at. It's not supposed to sound scripted, or like someone reading from a book. It's supposed to sound like a dinner conversation. Susan Stamberg once described good radio as akin to the guilty pleasure of listening in on a really interesting conversation at the next table in a restaurant.

Radio also gives people a chance to use their imaginations. Take the interview I did with Harold Varmus when he took over as director of the National Institutes of Health in 1993. I wanted to present Varmus as the academic scientist who didn't give a damn about the norms of Washington bureaucracy. So I interviewed him on his way to work, not in the government car that most agency heads used, but the way he always commuted: on his bicycle. You didn't have to see Varmus pedaling through traffic; all you needed for the mental picture was the up-close sound of traffic and a bicycle chain gliding through a derailleur.

Writing for broadcast comes in various flavors. I've written stories as short as 30 seconds, and as long as 30 minutes. Although it's rarer these days, the one-hour radio documentary is not unheard of. But in all broadcast formats, long or short, there's one crucial rule: Keep it moving forward. Your viewers or listeners can't flip back to the start to remind themselves what happened five minutes ago. If too much time has passed since you last introduced a character, introduce him again.

The best writing for broadcast, both radio and television, involves telling a story. Stories are engaging. They give you a structure. They have a beginning, a middle, and an end. They have characters. They set up a conflict, which helps you see a scientific issue in a more exciting way.

In radio, reporters sometimes fall into the trap of becoming beguiled by sound and forgetting about story. Tell the story. Let the sound help you tell the story. In fact, more often than not, you're not going to get great sound for a story. If you can get interesting sound in a molecular biology laboratory, you're a better man than I am. All I get is the white noise of refrigerators or fume hoods.

So be sure to get interesting interviews. Of course, that's harder than you'd think. Scientists are notoriously fond of jargon. Getting them to stop using it can be next to impossible. You can always try the approach my colleague John Nielsen uses. Tell your subjects to pretend that you are a potential funder, that you're drunk, and that you haven't got the faintest idea what their work is about, but they won't get a penny unless they can explain to you what they're doing.

Writing for broadcasting is, of course, writing for speaking. I hear the words I'm typing; sometimes, in fact, I move my lips when I write. There are obvious tongue twisters to avoid, of the "she sells seashells" variety, but there are other, less obvious phrases that you won't know cause trouble until you try saying them out loud. If you're writing for your own mouth, you'll just have to experiment to find what works and what doesn't. I've learned, for instance, that I have a hard time saying "researchers determined"—yet through some masochistic tendency I keep writing that phrase into my scripts.

The other issue you have to contend with is lung capacity. You may be able to write beautiful descriptions of the fine structure of the rocks on Mars. But if your sentences are too long, all that people will hear will be you gasping for breath.

Keep your sentences no longer than 10 words each. It's really not that hard. People do it all the time in normal speech. I'm doing it right now. Don't hesitate to use sentence fragments. Use action verbs. Best of all, verbs that allow you to omit adverbs. How about "ambled" for "walked slowly?" Or "shuffled"? Or "plodded"?

The typical radio piece alternates narration and sound bites, or *actualities*. I usually pick my sound bites first, and then write around them, always keeping the basic flow of the story in mind.

If you decide to try producing your own radio pieces, you'll need some basic equipment. Although audio engineers tend to sneer, consumer-grade minidiscs produce perfectly acceptable sound, especially if you are simply recording an interview, or getting the sounds of a research laboratory or telescope dome. A good omnidirectional microphone is a must.

Editing audio used to be a bloody business, using razor blades and splicing tape to physically cut magnetic tape and splice it back together. No more. Digital audio editing is a dream. If your ethics allow you to do so, you can clean up quotes, removing "ers," "ahs," and off-point dependent clauses to your heart's content.

Digital audio editing also allows you to adjust sound, add ambient sound, and fade one piece of audio under another, all things that used to require a fully equipped audio studio.

In television, as in radio, my colleague Peggy Girshman got the same question when she worked for NBC that I often get at NPR: "Oh, so you're not a writer anymore?" Writing is a part of the task, she says, but, with rare exceptions, no one just writes. Most people who write for television are producers or on-air reporters, although documentary units employ researchers and assistant producers who write as part of their jobs, usually to help the producer.

These days, with the quality of home video cameras, it's possible to consider making your own documentaries. Digital editing on home computers allows even novices to create reasonable video reports. They might not air as produced, but they certainly showcase your talents and might get you started on a new career path.

Just as radio depends on sounds, television depends on pictures. But here's a surprise that Peggy has pointed out: It also depends on anecdote. Not as in "It would be nice if I had an anecdote to start my story," more like "I have to have something on the screen to tell my story with." How this principle is applied depends on what kind of story you're producing.

Science on television generally falls into two categories: the short news-style piece, one and a half to three minutes in length, or the documentary, usually 30 to 60 minutes long.

The short newsy piece is often pegged to scientific journal reports, government announcements, or disease outbreaks. These pieces run on local or network nightly news programs or cable news weekly programs.

A typical story is one that was based on a study published in the *New England Journal of Medicine* that concluded that the less expensive streptokinase is superior to another clot-dissolving agent, tPA, in stopping a heart attack in progress. To produce this story, you would need to imagine the arc of it. Here's how Peggy pictures it:

First, summary by anchorperson of the news, introduces reporter. Next, reporter (that is, you) in voice-over of a picture. And the picture? Almost always it is a patient. After all, this is a story, and you want to engage a viewer who has a remote control at the ready and can change the channel in a fraction of a second. It would be impossible to start with the study results—people need a context of what the treatment does. It would be dull, and possibly confusing, to start with a picture of a blood vessel and start talking about the study results. So you have to find a patient to illustrate the point. Usually, that's done by contacting a local hospital or the doctors involved in the study.

You interview the patient on camera. Ideally, the patient is still in the hospital, having just survived a heart attack, so the experience is fresh. Less ideally, the shot is taken in a patient's home. Look for something interesting the patient is doing to avoid the dreaded "walking shot" (patient walks down the hall for no apparent reason). While sometimes a producer walks and talks with someone, there is always a need for a sit-down interview, which usually involves a quiet place and extra light.

Another essential element of the piece would be the expert interview, preferably with the author of the study. Sometimes medical journals or public relations firms put out an interview with the authors on a video news release (VNR), which can include pictures of the research lab or other relevant "b-roll" (for example, pictures of pills being placed in a pharmacy bottle). For producers without the resources to conduct their own interviews and shoot their own pictures, these VNRs can be very helpful, provided you keep in mind that they are undoubtedly presenting the most positive spin imaginable. Sometimes these VNRs are trying to create news where none exists and should be ignored. Depending on the rules of your news organization, you may be able to use these pictures or interviews as long as you can provide editorial balance. You would probably also want to talk to a local doctor—usually the physician who treated the patient you've interviewed. In order to introduce the doctor in your script, you'll need the doctor doing something, or else prepare for more walking shots.

For studies with any controversy at all, you would need to find a researcher who disagrees. The most you would use from either of these researchers would be two 10-second sound clips. By the time you tell the patient's story, including

10 to 20 seconds of sound from the patient; explain the study results, using the researchers to add content; and wrap up the piece, you've maxed out on time.

Finally, you would need additional b-roll—of streptokinase, of tPA, of an emergency room, of pictures of a blood clot. Some of that may be available for free—either from the VNR or from societies such as the American Heart Association. Creative use of graphics (either from an organization or designed by a TV graphic artist) could include a chart summarizing the main results and an animated graphic of how clot-busting drugs work. All this to make a two-minute piece work.

In the scripting, you work with the pictures, not to describe them but as support for the script. When you're talking about how many people could use the drugs, for example, you could show footage from an emergency room. It's not beautiful, but you need something to complement the copy. You have to have some image on the screen the entire time.

Long form is a lot like the short news pieces, in that you still have to find anecdote. But there is a much heavier burden: longer "scenes," to develop specific characters.

One television producer, Joe Blatt, doing a documentary for *NOVA* about searching for the top quark, had a hefty task. The scientific concepts he wanted to illustrate were all completely invisible to the eye—not an automatic winner for television. So Joe had to find gregarious, articulate, and—most important—significant players in the field. His budget (as with many documentaries) was 15 to 20 "shooting days," that is, days when he could use a crew. He had to plan carefully, to ensure that he would be able to shoot whatever action was possible. So, when an experiment was being conducted at Fermilab, Joe was there with a crew in the control room as the particles were being accelerated. There were many shots of people in front of computers, but they were talking and lively and engaged in the process of doing science. This program had a very heavy graphics budget, to help illustrate the concepts.

To develop the character of the scientists, sometimes you have to show the softer side of them—show them drinking beer at a bar with their colleagues outside of work, for example.

Only after all the pictures, scenes, and interviews are gathered can the actual writing begin. But much of the writing has to live in the producer's head all along: *What kind of scenes will I need to illustrate key aspects of the story? How can I get the best mix of folks talking, action happening, and my written script? How can I write that complicated science explanation in the 30 seconds that the graphics budget allows? How can I explain the science while still building character development so that the viewers can follow the story?*

It's a challenge, but one that is wonderfully satisfying once accomplished. No other medium, not even radio, has the capability to use all the senses. Viewers can get to know scientists, take advantage of moving graphics to understand complicated concepts, visit the middle of Antarctica or the Amazon jungle. It may not be writing in the classic author-in-the-lonely-study sense, but it is all about using all your skills to creatively capture the imagination and the mind of the viewer. Which is why when people ask me now whether I miss writing, I just smile and feel smug.

12

Freelance Writing

KATHRYN BROWN

Kathryn Brown recently launched a specialty communications company—EndPoint Creative, LLC—that offers writing and editing services to science, medical, and technology organizations. Previously, as an award-winning freelance journalist for 12 years, she contributed to *Science, Scientific American, Discover, Popular Science, Technology Review,* and *New Scientist,* among other magazines. Kathryn is a former board member of the National Association of Science Writers, former chair of its freelance committee, and recipient, in 1999, of NASW's Everett Clark/Seth Payne Award for Young Science Journalists, among other awards.

Living on the outskirts of Washington, D.C., I often encounter that classic cocktail party question, "And what do you do?" When I say I'm a freelance writer, people tend to respond in one of two ways. Some lean forward, suck in their breath, and marvel at how romantic and free my days must be. Others cluck, tilt their heads, and mutter that life must sure get bleak, scratching out a living on words.

The truth lies somewhere between.

Among freelancing's perks, flexibility ranks high. I might write about biodefense one day and mental health the next. I might work in the library, on the porch—or, yes, in my gym clothes. My time is my own. The price for this luxury? Responsibility. Freelances are entrepreneurs. It's our job to find work, negotiate that work, and handle all the business details, from taxes, insurance, and retirement accounts to business cards and printer paper.

If you're starting—or nurturing—a freelance science writing career, here are some practical tips to consider along the way.

Diversify

From cloning and stem cells to space exploration, food, and the environment, the visibility of science in society is strong. That's good news for freelances. Don't limit yourself to a handful of poorly paying publications, when you might be growing and learning (not to mention earning a better living) by diversifying. Today's successful freelances often go beyond writing straight magazine articles to writing trade books or projects for nonprofits, Web Sites, corporations, government agencies, and public relations agencies.

As you diversify, hunt for one or more anchor clients—those who will hire you repeatedly. Also, consider following the tried-and-true technique of trading up. If you need experience, there's nothing wrong with freelancing for a city paper or tiny nonprofit (not big money-makers). Afterward, you can build on that experience—returning to those clients, while also using your clips to pitch other, bigger projects elsewhere.

Diversifying doesn't mean you have to write everything. What if you have a passion for one field, in particular—say, physics? Just as staff writers cover a beat, some freelances feel most comfortable specializing in something they can learn top to bottom. And that can be done. One solution is to build a diverse freelance clientele—for instance, blending work for consumer magazines, trade publications, books, or associations—in a single subject area.

Learn to Negotiate

Freelance writers often share a "David versus Goliath" view of publishing. (Guess which one we represent.) But top writers across fields say that deal-making need not be so grim. Before you sign a contract or agree to the terms of an assignment, think for a minute. What are your goals? Do you need something more—more money, more time, more rights? If so, ask. If you're good enough to do the work, you're good enough to negotiate the deal. For inspiration, look for workshops or publications on making business deals.

Manage Your Money

The very best clients pay well and promptly . . . and then there's everybody else. Whether you're writing for a magazine, nonprofit, or PR agency, chances are you'll be eagerly awaiting your check for 30 days (or more). While you wait, however, your debts won't—from quarterly taxes to the telephone company, expenses pile up. That's why you need a budget. How much money do you spend each month? How much do you bring in? Either on your own or with an accountant, you need to evaluate income and expenses to be sure the cash keeps flowing.

Write Well

It's obvious, yet it bears repeating: When you're a freelance writer, every assignment counts. Don't get careless. Over the past 12 years, I've learned how to pace my freelance work schedule. But there have been painful reality checks—and the weary midnight work session—along the way. Keep a calendar close at hand, and always ask for more time up front, if you think you'll need it to get an assignment done well.

Making It Work

Some people try freelancing for a little while, only to race back to an office job. They often cite the same reason: isolation. Working on staff is like being a team player. Freelancing, however, is an individual sport. Some people are suited to long hours of unstructured quiet—others quickly come to resemble Jack Nicholson in *The Shining*. If you're considering freelancing, ask yourself whether you're comfortable working alone for hours (or days, weeks, months).

Even reclusive freelances need to break out of the basement. Often. For one thing, it's good business. You need to network with colleagues, either by having lunch or by attending the events of local professional groups. (See NASW's website, www.nasw.org, for a list of local science writing groups.) If you hate schmoozing, make yourself go just once a month. Even if you don't swap a single business card, you'll share stories and laughs with peers, which makes the daytime silence more bearable. If, on the other hand, you love to socialize, roll up your sleeves and get involved—every organization needs volunteers.

You might also learn lifestyle tips. Everybody has a different way of making freelance life work. Some writers find it best to get dressed, get out of the house, and work in a rented office. Others prefer to shuffle downstairs in a bathrobe, coffee mug in hand, to greet the computer screen. You might find the perfect method to your madness in someone else's routine.

Which brings me to a final point. With fewer meetings to attend and colleagues to chat with, freelances sometimes get more done in a day than their peers on staff. Play to those strengths. If you've completed two dozen creative projects in the past year, for instance, craft a résumé that flaunts them. If you're a generalist, consider giving editors a list of your recent publications, broken down by field, along with your next story pitch. If you're a specialist, draft a work summary that illustrates your depth of expertise. In the end, every freelance brings unique talents to the table—and if you do your job well, those talents will shine.

RESOURCES

American Society of Journalists and Authors *Contracts Watch* newsletter: www.asja.org/cw/cw.php

National Writers Union: www.nwu.org

NASW Freelance Website: www.nasw.org/mem-maint/freelance/ *The ASJA Guide to Freelance Writing,* Tim Harper, ed. (New York: St. Martin's Griffin Paperback, 2003).

13

Science Books

CARL ZIMMER

Carl Zimmer divides his writing time among science books, magazine and newspaper articles, and a Web log. His articles appear in publications including the *New York Times*, *National Geographic*, *Newsweek*, *Popular Science*, *Discover*, *Natural History*, and *Science*. His books include *Soul Made Flesh* (2004), *Evolution: The Triumph of an Idea* (2001), *Parasite Rex* (2000), and *At the Water's Edge* (1998). He also writes "The Loom," a Web log about science that receives 7,000 visits a week (www.corante.com/loom). Carl graduated magna cum laude from Yale with a B.A. in English in 1987. He joined the staff of Discover in 1989, where from 1994 to 1999 he was a senior editor. In 2002 he was named a John Simon Guggenheim Fellow. He has been awarded the Pan-American Health Organization Award for Excellence in International Health Reporting and the American Institute of Biological Sciences Media Award.

Every piece of science writing has a trajectory, a life history. You decide you want to write something, you find a subject to write about, you find someone to publish it, you research it, you write it, and then—if all goes well—it eventually turns up in print. These milestones mark the life history of every piece of science writing, whether it's a magazine feature, a newswire story, a post on a Web log, or a book. But these genres are a bit like animals. Every species has its own life history. All animals are born, grow, and reach maturity, but each species takes its own route from one milestone to the next. You can't equate the life of a mayfly with the life of a tortoise. Here, then, is the life history of a science book. I hope that in describing it, I convince you that the science book is not simply a very long article, but an altogether separate beast.

The Idea

Book ideas come about in many different ways. The idea for my first book, *At the Water's Edge*, occurred to me one day in 1996 as I was sitting at my desk at *Discover*. I had just written an article about how our fish-like ancestors crawled on land 360 million years ago. I was flipping through the published article, sometimes glancing up at my stack of notes and papers. The stack was a foot thick. (I'm very slow about clearing off my desk.) I probably managed to get half an inch of that information into the story. All the rest of those wonderful stories within the story—about the evolutionary principles these animals illustrate, about the 150 years of scientific debate over this central mystery of our heritage—would never see the light of day. I thought about the other articles I had written on other great evolutionary transformations, and all the details I had left out of them because of space constraints. I decided to write a book.

I suspect this is a common route to many first science books. Others are born when authors are approached to write a companion book to a television series, or to serve as a co-author with a scientist. Or a writer may simply wake up one morning and realize that a great book on subject X has yet to be written. How ever you come up with your book idea, you should pause a moment to savor the prospect of someday seeing your book on the new nonfiction table at the bookstore.

And then you have to decide: Can you really bear to write this book?

The book business can be almost absurdly brutal. Only a minuscule fraction of book proposals are accepted by publishers. Most of the advances paid, especially for science books, are less than $50,000—which is meant to cover at least one year, or sometimes two or three, of fulltime work. After all that, even if you get a book contract, even if you can survive on the advance, even if your book gets published, it will still be just one out of some 195,000 titles published every year.

While these are bleak numbers, they are not cause for total despair. Science books rarely storm the citadel of bestselling nonfiction, which is filled mostly with books about politics, history, and professional wrestling. Once in a blue moon a book like Jared Diamond's *Guns, Germs, and Steel* (1997) or Brian Greene's *The Elegant Universe* (2003) manages to hit the big time, and for years to come, book agents and editor and writers try to re-create their magic. ("It will do for alternating current what *Longitude* did for clocks!") But a fair number of science books manage to thrive just below the bestseller radar. Generally, these successful science books offer a great narrative, compelling arguments, or a surprising new way of looking at the world. Some even have all three. Ask yourself if your idea has any.

If, after taking all this into consideration, you still have not abandoned your idea, you need to put it to a final test before writing a proposal: Do you love it?

Love is crucial to writing books for three reasons. First, readers can tell when an author's heart isn't in the writing. Second, the book will be a dominant part of your life—from proposal to manuscript to publication—for at least a couple of years. If you don't love it, you will eventually come to hate it and, by extension, yourself. And finally, there is the possibility that once your book comes out, you may well watch it slip into obscurity. You need to love the idea enough to be content that the book simply exists.

The Proposal

If your idea passes these tests, it's time to write a proposal. Successful proposals vary wildly. Some are 10 pages long, some a hundred. But most share certain ingredients. They show that the author has planned out the shape of the entire narrative. Obviously, you can't know all the details before you've fully researched the book, but an editor needs to feel secure in your hands. The proposal also needs to demonstrate that you can write a book. This is especially important if you've never written a book before. Books have a symphonic structure, with smaller story arcs riding on top of bigger ones, with digressions and flashbacks that come together into a logical resolution several chapters later. An extended outline of the book or a sample chapter will persuade an editor that you can meet this challenge.

Writing a proposal is more than just a way to snag a contract, though. The more work you do on a proposal, the less you have to do on the book itself. It's important, for example, to make sure you can find enough to talk about in 75,000 to 100,000 words, but not too much. An idea that would make for a good magazine article will be diluted beyond recognition as a book. On the other hand, if you pick an outrageously vast topic—the complete history of space flight, for example, or a survey of every major infectious disease—you may wind up spending 10 years writing a multivolume opus.

I nearly fell into this trap myself while planning *At the Water's Edge.* Initially I had wanted to write 10 chapters, each of which would look at an evolutionary transition—fish coming onto land, birds moving from ground to air, and so on. When I realized how ridiculously long my book would be, I did some radical surgery. In my proposal, I laid out a plan to write about only two transitions: fish coming onto land, and land mammals going back to the water as whales. I'm glad I was willing to surrender my grander ambitions—*At the Water's Edge* wound up much longer than I had expected, even with its modest scope.

When you write the proposal you should also think about the structure of the book. Ask yourself what sort of narrative will drive the story best. Some great science books are structured as a race—a recent example is James Shreeve's *The Genome War* (2004), which followed public and private researchers trying to finish the Human Genome Project first. Other books, such as John McPhee's *Basin and Range* (1982), are intricate meditations. McPhee drives cross-country with a geologist and weaves history and geology into the trip. Others are epics, like Richard Rhodes's *The Making of the Atomic Bomb* (1987), which tracks a staggering cast of characters over many years. You may feel most comfortable using one of these structures; on the other hand, the subject matter may demand a structure of its own.

The Contract

Once you have a proposal in decent shape, I recommend finding a literary agent. You may balk at giving up 15 percent of your earnings, but good agents earn every penny of their commission. To find a good agent, look at science books that are similar to the one you want to write. Usually authors thank their agents in the acknowledgments. You can find the addresses of these beloved agents in the latest edition of Literary Marketplace.

If all goes well, you will find an agent, and your agent will find a publisher (or several) ready to negotiate a contract for your book. While your agent can be a great help in these negotiations, you need to have a solid understanding yourself of what should and should not go into a contract. Contract matters are complicated and are outside the scope of this discussion; to learn more about them, see the Authors Guild website (www.authorsguild.org).

Research

If you've put enough work into your proposal, figuring out the research required for your book should be relatively straightforward. It won't be easy or quick, but at least you'll know what you have to do.

Many of the skills involved in researching shorter forms of science writing carry over to books. In fact, books are so similar to magazine articles in this respect that it's sometimes possible to profitably piggyback book research on magazine assignments. (Be careful, though—your magazine editor may not like the idea of subsidizing book work, and you may not want to give away your most exciting material in an article that comes out long before your book.) For books, just as for articles, I spend a lot of time paying visits to scientists, watching them do their work, interviewing them at length, and working through the transcripts in order to create narratives of their discoveries.

In my experience, the biggest difference between researching books and researching articles comes with reading. When I'm writing a book, I spend lots of time in libraries, searching out old books, journal articles, letters, notebooks, and other documents that help me reconstruct how Aristotle influenced Darwin, for example, or how alchemy helped give rise to a science of the brain. This research may require you to travel to special library collections, which can be very time-consuming. But even in the past few years, computer databases have made historical research much easier. For my most recent book, *Soul Made Flesh*, I traveled to England in the spring of 2002 to look at seventeenth-century books and pamphlets at the British Library and the Bodleian Library at Oxford. If I were to do the same research today, I wouldn't even have to leave my desk. A database called Early English Books Online now contains *every* book published in English before 1700. That's progress.

Writing the Book

Once you set about starting to write the book itself, you will see how good your proposal was. It may turn out that your research is tugging you in a different direction than you had planned. You shouldn't ignore that tug, but you also should not let yourself get tugged every which way. I sometimes find when I'm deep into writing a book that I need to research a chapter's worth of new material. It's fine to take a break from writing to do some more research; you can return to your manuscript with fresh eyes.

It's best, I believe, to hold off on starting to write the book until you've done most of your research. You don't have to have hunted down every detail first, but it's a good idea to feel in control of the subject before you try to shape it into a narrative.

The actual writing of a science book poses some special logistical challenges. You're dealing with a mountain of information that's too big to keep in your head at once. Find a way to organize it. I like visual displays of information, so I'll write out my book outline on giant pieces of paper that I tack on a wall. I'll note along the way how my various pieces of research will fit together into a single narrative, with lots of arrows pointing forward and back through the outline. It's also a good idea to keep track of your sources with a software program such as EndNote. It will let you assemble footnotes and a bibliography with relatively little pain.

In the last stages of writing, you need to weed out the many mistakes—both factual and stylistic—that have inevitably sprouted across your pages. You, not your publisher, are your own fact checker. Try to show portions of the manuscript to as many sources as you can. Each one will find a different set of mistakes.

To catch stylistic mistakes, I show the manuscript to my wife, a perceptive reader and excellent editor. Other writers read their manuscripts aloud into a recorder and then listen back to the tape. They've got a stronger constitution than I have, I'll admit.

Finishing

When I turned in the manuscript of *At the Water's Edge* to my publisher, I felt as if I had created a perfect object. My work was done. This delusion, I have since learned, is common among first-time authors. It would be as if a couple decided that after two years of parenting, their child was ready to get an apartment of her own. You must shake off this delusion as fast as you can if you want your book to fare well, because much work remains.

First comes the editorial letter. Your editor will send back the manuscript to you with a note that begins with something like, "Dear Carl: Thank you so much for your wonderful manuscript. It is a delight. There are just a few parts that need some tightening up . . ."

Then will follow several pages of massive changes, or what your editor will probably call "tweaking." You may have to destroy entire chapters. Other chapters may need to run in reverse. And when you turn from the editorial letter to the manuscript itself, you find it covered in red-ink chicken scratch.

This is good. This is your editor doing what editors should do. Your editor may be wrong in some cases, and you should resist changes that would diminish your book. But editors are right far more often than authors would like to admit. You should spend several weeks seriously addressing your editor's concerns.

Even at this point, you're not finished. The copy editor now has a crack at the manuscript, combing through it for inconsistencies, bad grammar, and fuzzy language. Then comes the proofreader. The manuscript may go through five or six versions by the time it finally gets turned into a real book between covers. You will reread the book until you are sick of every word, and then you will have to do it several times more. (This is where loving your subject can help you avoid going insane.)

And even as the editing tapers off, more work comes to take its place. Editors have to write catalogue copy, back jacket copy, flap copy. They have to send the manuscript out to garner blurbs. They have to meet with marketing people to figure out how to pitch the book to booksellers, and with publicists to figure out how to get the word out. Your editor needs your help at every stage of the way, even if this means taking time away from wrapping up the manuscript to fill out a lengthy questionnaire. The more help you can give, the more likely it is your book will thrive.

Selling the Book

This may not seem like part of your job as a book author, but these days it needs to be. As a science writer, you have a mental database of publicity opportunities that dwarfs anything your publicist can offer. Perhaps six months before your book comes out, you should talk to your publicist to get a sense of what she plans to do, and then figure out how you can complement her work. (Doing nothing but giving your publicist a hard time is not going to help your book.) For your own contribution, think hard about where you can give talks—anywhere, from museums to churches. Investigate online communities that might be interested in your book (cancer support groups, futurists, etc.). Set up a website if you don't already have one. If you start working on magazine articles that will come out around your publication date, try to think of ideas that will attract people to your book, which you should mention prominently in your contributor description. Every day, think of something you can do to promote the book—at least until you start working on your next one.

14

Popular Audiences on the Web

ALAN BOYLE

Alan Boyle is science editor for MSNBC on the Internet. His first job in daily journalism was on the graveyard copy desk on the *Cincinnati Post,* and after that he held a variety of editing positions at the *Spokesman-Review* in Spokane, Washington, and the *Seattle Post-Intelligencer*—gradually working his way from the editor's side of the desk to the writer's side. As foreign editor at the *P-I* in the 1990s, he helped organize seminars on online media for Russian journalists and wrote an online journalism guide for Asia-Pacific journalists as part of a UNESCO program. In 1996, he joined the "launch team" for MSNBC.com and soon settled into the space and science beat. In 2002, his report on genetic genealogy won NASW's Science-in-Society Award and the AAAS Science Journalism Award in the online category.

Let's face it: We're all Web journalists now.

You might be working for a newspaper or magazine, a television or radio outlet, but your story is still likely to end up on the Web as well as in its original medium. You or your publication may even provide supplemental material that appears only on the Web—say, a behind-the-scenes notebook, an interactive graphic, or a *b*log.

Or you might even be a journalist whose work appears almost exclusively on the Web—like me.

I worked at daily newspapers for 19 years before joining MSNBC, a combined Web/television news organization. So I still tend to think of the Web as an online newspaper, with a lot of text, some pictures, and a few extra twists. But with the passage of time, online journalism is gradually coming into its own—just as TV started out as radio with pictures, but soon became a distinct news medium.

To my mind, the principles of online journalism having to do with fairness, accuracy, and completeness—are the same as the principles of off-line journalism. But the medium does shape the message, as well as the qualities that each medium considers most important. Wire-service reporters value getting the story out fast; newspapers value exclusive sources; magazines value in-depth coverage; radio and TV look for sounds and pictures that will help tell the story. All these factors are important for the Web as well, but one thing makes online journalism unique: Web writers are looking for ways to tell the story using software.

Let's take a closer look at how one multimedia story unfolded, then get into how the tools and toys of the trade can be used in your own work.

Case Study: The *Columbia* Tragedy

News coverage of space shuttle launches and landings usually follows a familiar routine: From MSNBC's West Coast newsroom in Redmond, Washington, I would update the landing-day story continuously, starting with the de-orbit burn, just as a wire service reporter might do.

On February 1, 2003, however, the shuttle's landing was scheduled for a Saturday morning, one of the lightest times of the week for Web traffic. So I departed a little from the usual script: I decided to sleep in, and let the East Coast news desk handle *Columbia* updates.

When the phone woke me up a little after 6 a.m. Pacific time, I wasn't thinking about *Columbia* at all. But when projects editor Mike Moran told me the shuttle was missing, it took just a few seconds to realize that something terrible was happening. By the time I arrived at the newsroom, a couple of extra editor/producers were already lending a hand to the weekend's skeleton crew, and artists were on their way in.

Also arriving were the interactive producers, or IPs, whose role is unique to Web journalism. IPs combine text and images, audio and video, and turn it into computer code. The resulting interactives help tell the story or explain the issues in Web-friendly ways.

While Mike continued to keep up with the breaking news, I helped get a new crop of interactives started. One was a clickable graphic showing the stages of the shuttle re-entry process, and another was a guide to the shuttles and their components. We also updated our mini-biographies of the crew and our timelines of shuttle flights to reflect the new tragedy, and selected NBC videos to accompany the story as it evolved.

After getting the interactives started, I turned to the lead story—and as I kept track of the day's developments, I also started thinking about the tragedy's implications for the entire space program. What would happen to the international space station, and all the plans for multimillion-dollar space tourism and entertainment ventures? It was clear that NASA would have to rely upon the Russians for space station support, and that there would be no pop stars or millionaires in space for a long time. Between the rewrites, I gathered information for a sidebar about the space station's future.

The editors in our West Coast newsroom told MSNBC television producers on the East Coast that I was working on the space station angle—and that struck a chord with the TV operation. So a producer called me up, did a pre-interview, and arranged three TV "hits" on the subject, going as late as 11:30 p.m. Pacific time.

As the tragedy unfolded, MSNBC received hundreds of e-mails from users asking questions about the tragedy as well as the shuttle program. I put together a Q&A addressing the most common questions, and helped the Opinions editor set up a mailbox for *Columbia* condolences.

By the time Saturday turned into Sunday, I had turned the lead story over to a writer/editor on the evening shift, and fine-tuned the interactives and sidebars that had been created throughout the day. I had the beginnings of the space station sidebar, but that story wouldn't take final shape until a couple of days later. The day's final task was to update my Web log with links to our "In Memoriam" mailbox and other online condolence books.

"It's been a long 19 hours," I wrote in a posting time-stamped 1:23 a.m. Pacific time Sunday. "For some reason I feel there's little more I can add at this point, other than to send condolences and prayers to the families of the fallen and to the extended NASA family."

There would be much more to add in the days that followed, of course. It wasn't my first 19-hour workday on the Web, and it wouldn't be the last.

Writing for the Web

Web stories have to be shorter and snappier than stories in print. That's certainly a challenge for science writers, since the details of a scientific discovery can be so important. But there are ways to make even a complicated story more palatable for click-happy Web users:

Be Direct

Wire-service leads tend to work better than the indirect, magazine-style lead. That's doesn't rule out adding a little drama to the start of your story. In fact, the story that brought me an NASW award for online journalism ("DNA Takes on a Family's Mysteries," msnbc.msn.com/id/3077144) began with what could be called an indirect lead:

> REDMOND, Wash.—I clutched the phone and started the trans-Atlantic countdown: Thanks to a mail-order DNA test, I was about to find out whether my Irish "cousin" was really my cousin. On the other end of the call was my cousin's fiancée, who read off 10 numbers while I compared them. The first number? Check. The second? Check. So far, so good.
>
> My quest was coming to a climax after four years of researching my Boyle family tree. . . .

I tried to give a general sense of what the story was about in the first paragraph, and lay out the full thrust of the story in the first four paragraphs. The point is that you have to get to the point quickly in a Web article—which I suppose is a good rule to live by for other media as well.

Chunkify

At the MSNBC website, stories can run as long as 1,200 words before the editors squawk. In those cases, however, we try to organize the story into modules or "chunks" of 300 to 600 words, separated by subheadings. That improves readability on the Web, because the story doesn't look so daunting, and readers can skim past modules if they want to. It's also good discipline for writers. So sharpen your outlining skills and break that story into smaller chunks.

Modular organization makes it easier to deal with technical details of a story that may be of interest to some readers but not to others. For example, during the heyday of HotWired.com, Simson Garfinkel generally stuck a "Geek This" switch in his "Packet" column on tech topics. When you clicked the switch, the geeky details popped in right at the appropriate place in the column. Clicking the switch again zipped the column back into its abridged version (hotwired.wired.com/packet/garfinkel/97/20/index2a.html).

Accessorize

It's a short hop from the modular approach to an approach in which the supporting elements of a story become clickable boxes and sidebars within a story. Flash-based graphics, blogs, forums, chats, and other "accessories" could even be used as the primary media for telling the story. During the fall of Russia's Mir space station, we posted updates to an MSNBC Web forum before we incorporated them into the story.

Hyperlinkify

Unlike a magazine, newspaper, or broadcast network, online stories can provide hyperlinks to resources elsewhere on the Web. Thus, instead of having to explain in depth what a polymerase chain reaction is, you can provide the ungeeky, shorthand explanation and link to an authoritative explanation for readers who need more background. The effect can give you the kind of power that Woody Allen fantasized about in the film *Annie Hall*, when he pulled the real Marshall McLuhan out of a corner to settle an argument about McLuhan's views on media.

A Toy Maker's Toolbox

Over the years, Web journalists have developed a standard set of techniques for going beyond static text, pictures, audio, and video of older media. Here are some of the frequently used tools and toys of the trade:

Galleries

A slideshow is a series of pictures and captions you click through, one by one. The captions can be enhanced by audio or video. A video gallery is a variation on the theme, where you can click on a series of video highlights. An example in the MSNBC Special Coverage section is "The Week in Pictures" (www.msnbc.msn.com/id/3842331/).

Pop-Ups

Self-contained, Flash-based graphic presentations can be created to provide a virtual tour, explain how something works, or tell a story. For some examples, check out the science interactives listed on the "Cosmic Log Links" page (family.boyle.net/spj/) or try out MSNBC's "Fueling the Future" simulation of energy policy choices (www.msnbc.com/modules/fuel_future/game/).

Surveys

Our "Live Vote" asks users what they think about a controversial subject: *Do you think stem cell research should be banned? Do aliens exist?* But features like this should never be regarded as valid public opinion surveys, and as a science writer, you probably already know why—primarily because of the self-selecting nature of the survey. They're purely for fun, just another toy.

Quizzes

These are multiple-choice quizzes that are scored by the computer when you hit a button (usually using Javascript). You can use them for brainteasers or for more serious purposes—for example, a quick assessment of an individual's heart health risks, which we feature on MSNBC (www.msnbc.com/modules/quizzes/heart_dw.asp).

In-Story Primers

Large masses of facts—ranging from glossaries to timelines to profiles of key players—can be handled in a scrollable or clickable space embedded within the text. For an example, take a look at the shuttle fact file and the astronaut profiles in one of MSNBC's stories about the *Columbia* disaster (www.msnbc.msn.com/id/4088826/).

Forums

Web-based discussion boards are probably the most interactive features you can offer. They let users talk about the news with each other as well as with experts who are monitoring the discussions. But they tend to be high maintenance: The people in charge have to rein in the blowhards, trash-talkers, liars, and spammers. The Habitable Zone (www.habitablezone.com) is an example of a well-behaved Web community. At MSNBC, we tend to go for a mailbag, which is an online "letters to the editor" offering (www.msnbc.msn.com/id/4045141/), or a chat, which may be structured like an Internet Relay Chat or like an audio talk show with questions submitted to an online moderator (chat.msnbc.com).

Blogs

These Arrays of time-stamped Web postings can serve as a reporter's journal, a briefs column, a means of stimulating and organizing feedback to stories, or an ephemeral guide to websites of interest (www.cosmiclog.com). But they also can be applied to breaking news situations: Spaceflight Now, for example, uses the blog technique (a.k.a. "Mission Status Center") to great advantage in covering rocket launches, Mars rover landings, and the *Columbia* tragedy (www.spaceflightnow.com/shuttle/sts107/status.html).

■ ■ ■

In short, science stories on the Web can offer features you won't find in any other kind of science story. Here are just a few examples:

- A slide show that displays the Hubble Space Telescope's greatest hits and invites users to vote for their favorite
- A clickable graphic that explains how NASA's Mars rovers work and shows where they're going and why
- A quiz that helps users size themselves up against lifestyle-related health risks
- A moderated bulletin board where users can register their feedback on health issues
- An interactive "documentary" about 9/11 that blends sights, sounds, and databases explaining how and why the World Trade Center towers fell, who was responsible, and who was touched by the tragedy

And that's the great thing about science writing for the Web: You're not locked into the print paradigm for telling the story. There are rules, but they're new rules, and they're constantly changing. Web writing is probably as close as journalism gets to toy making—that's how much fun it can be.

15

Science Audiences on the Web

TABITHA M. POWLEDGE

Tabitha M. Powledge has written about science and medicine for more than 30 years. After receiving an M.S. in genetics from Sarah Lawrence, she worked at the Hastings Center, a bioethics think tank. Her work has appeared in *Scientific American*, the *Washington Post*, the *New York Times*, *Current Biology*, *PLoS Biology*, *The Scientist*, *BioScience*, *Archaeology*, *Health*, *Popular Science*, *Nature Medicine*, and *The Lancet*, among others. Tammy was senior editor of *Nature Biotechnology* and founding editor of *The Scientist* and is now a contributing editor to the National Academy of Science's *Issues in Science and Technology*. She is the author of *Your Brain: How You Got It and How It Works* (1994). Tammy has twice been awarded fellowships from the Knight Center for Specialized Journalism and has received the Distinguished Communication award of the Society for Technical Communication. She has written for several online publications for scientists, among them *BioMedNet*, *HMS Beagle*, *Genome News Network*, *SAGE KE*, and *The Scientist*.

In otherwise hard times, at least one market for science writing appears to be expanding: writing for scientists, particularly online. It's also a market that can offer unusual professional satisfaction. When you write for scientists, you can ignore many of science and medical journalism's topical fads. On the Web, you can pursue subjects that interest you, delve into more of their technical details, and write about them with surprising flexibility and freedom.

Like everything else in the dot-com world, online-only publications for scientists have come and gone. I, for one, am still mourning the disappearance of *BioMedNet,* which Elsevier dropped at the end of 2003. For several years *BMN* was an important market. It published at least a couple of news stories every weekday and also covered several basic research conferences annually.

But there's good news, too: A few online news operations allied with print publications are still going strong. These outlets, such as *TheScientist.com* (www.the-scientist.com) and *NewScientist.com* (www.newscientist.com), publish unique content that does not appear in their print versions. Top weekly journals also publish daily news online—among them *Nature* (www.nature.com/news) and *Science* (sciencenow.sciencemag.org). So does the top-tier publication *Scientific American* (www.sciam.com), which appeals both to those with an armchair interest in science and to scientists themselves. The stories in these online publications—typically short, in the range of 400 to 600 words—are written by both staffers and freelances.

One of the best things about writing for scientists on the Web is that it's not like typical Web writing at all. It resembles traditional print writing—but, amazingly, often with fewer constraints. And it is garnished only lightly with electronic doodads. Publications for scientists are not mad for multimedia, so your words don't have to take second (or third) place to video documentaries, interactive quizzes, Flash animation, or chat. Hyperlinks, yes, but only rarely will there be slideshows or snazzy static graphics.

Nor is this a deeply collaborative process. Usually it's just you and your editor, who often leaves you to produce your piece in your own way. This is different from Web writing in general, when you might be part of a Web content team whose other members regard you as the least valuable player.

In Web writing, as in print, when you write for scientists instead of the general public, there's more to write about. "For technical audiences, I delve deeper into how the findings are being received by others in the scientific community," says freelance Dan Ferber. "Rather than just a quick reaction quote, I might explore hurdles researchers face and how they might overcome them."

News writing for scientists is a growth area on the Web, which requires a writing style that is different from typical magazine-style feature writing. "I can't tell you how many successful magazine writers have had miserable experience writing news for me because they simply didn't pay attention to the style," says Ivan Oransky, who edits online daily news for TheScientist.com. "We attribute things more carefully and tend to write leads that include news, rather than interesting anecdotes or background." Ivan says his writers need to be able to cover a fire in a laboratory as well as they would cover the latest study published by that laboratory.

Whether you're writing news or features, you need to assume a slightly different tone for a technical online audience. There's no need to define your terms, or to explain some concepts basic to the workings of science, like the need to replicate research results. But this is a balance that some writers have trouble achieving, according to Christine Soares, former online news editor at TheScientist.com.

"Take as a completely random example the HapMap project" of the National Human Genome Research Institute, she says. "For a general audience, you're going to focus on basic details of what they're doing and what the ultimate application will be at the 'consumer' level. For a scientist audience, you're going to focus on the structure of the project and its funding, who's participating, as well as any new scientific techniques or procedures that will be employed."

Former *BioMedNet* editor Henry Nicholls points out that aiming for a scientific audience can greatly expand the types of pieces you write. "A lot of interesting stories presumably aren't covered in nonspecialist media because they are just too complex," he says.

You can often write short and fast for the Web. Online writing tends to be shorter than writing for print: shorter paragraphs, shorter pieces overall. The assumption is that readers have a hard time trying to absorb large quantities of text from a screen. But few people realize that you can also write very, very long. For those of us who like to stretch out, online publications—ones for scientists, anyway—often are more flexible about space. Sometimes considerably more flexible, hundreds of words more flexible.

I know, this is not what you've heard. And it's not true at most online consumer publications, where graphics rule, the editors assume that viewers read at a third-grade level, and words can be the least important part of the story. But scientists are accomplished readers—whether they know how to write, of course, is a different matter—and they tend to want detail. Lots of detail. They are thoroughly accustomed to plowing through the gnarly prose of their peers. Your writing will, by contrast, be a pleasure, even if it's scrolling by on a computer screen.

Readers tend to think of online publications the way they think of newspapers: instant articles, disposable content, a quick read. The publications tend to think of themselves that way too. This makes Web writing a good way to break into science writing. "There's a constant need for 'content,'" says Christine, who is now a print editor at *Scientific American*. Online "is an easier place to stick a toe in the water, learn some good habits if one has a good editor, and generate the first few clips," she says. "And for experienced writers, it's also a good place to turn unused threads from a larger project into a few quick stories that generate extra income."

16

Science Editing

MARIETTE DICHRISTINA

Mariette DiChristina joined *Scientific American* as executive editor in the spring of 2001. Previously, she served as executive editor at *Popular Science*, where she worked for nearly 14 years. Her work in writing and overseeing articles about space topics for that magazine was honored when the Space Foundation gave *Popular Science* its 2001 Douglas S. Morrow Public Outreach Award. Before coming to magazines, she was a reporter for the Gannett Westchester Newspapers, now known as the *Journal News*, and a stringer at several papers in New York and Massachusetts. Mariette is on the board of the National Association of Science Writers and is the former chair of the board of Science Writers in New York. She studied journalism at Boston University and lives in Westchester County, New York, with her husband and their two daughters.

Let's be honest. Editors, as any writer will tell you, aren't all that bright. They may say they're looking for stories that will teach something important about the way the world works, but mostly they want to be entertained. They can't follow leaps of logic. They get distracted by elaborate prose, and they have no patience for boring factual details. They get confused by too many characters in a narrative, or they're easily irritated by extraneous quotes. And they don't like big words very much, either.

In other words, we editors are a lot like the readers that we—and you—are trying to reach. In fact, we're a special kind of reader, in that our livelihood depends on our ability to think like the audience of our publications. This is the case for any kind of editing, not just science editing. Writers may shift tone or approach for different markets, but editors live and breathe our readers' way of life. We must internalize their interests, who they are, and what they expect from our magazines, newspapers, or Web Sites. Editors know what level of scientific language our readers will understand and what they won't. Each one of us also deeply understands our publication's unique mission.

Many people say that to be a good editor you first have to be a good writer and reporter. We editors like to think so, too. Having had experience as a writer helps inform good editing, and gives the editor a firmer appreciation of the reporter's point of view. And it's certainly true that, if necessary, an editor must be able to step in and complete the reporting and revisions on an article. But more than being good writers, editors must be good *critical thinkers* who can recognize and evaluate good writing—or can figure out how to make the most of not-so-good writing.

Especially when the subject is science, which can be complicated and convoluted, a good editor needs a sharp eye for detail. We need to be organized, able to envision a structure for an article when one does not yet exist, or to identify the missing pieces or gaps in logic that are needed to make everything hang together.

In a way, an editor's job, whether it's for a science publication or for any other kind, is a bit like that of a television producer. In addition to clarifying the story line, we're responsible for developing the entire "package" (to use a magazine-industry term). Toward that end, we coordinate not only with the writer, but also with layout designers, illustrators, photographers, and photo editors. In a glossy, illustration-rich publication like *Scientific American*, this producer function is especially important—and time-consuming.

Editors supervise and guide the stories from initial assignment through final proofs. We juggle articles in different editions: As we close one issue, we're beginning work on another and are midstream on a third. This can be tricky at a monthly like mine, downright maddening at a weekly like *Science* or on a daily newspaper's science desk.

Above all, though, I think of an editor's job as that of First Reader—the person who, ideally, will be the writer's sounding board, coach, on-staff advocate, and, ultimately, constructive critic, in pursuit of the mutual goal of turning a good story into a great one. How do editors perform these functions? Who are those people on the masthead? What are their specific tasks?

Maybe the easiest way for me to sketch some of the answers is to take you on a hypothetical editorial journey, from a feature article's proposal to its publication. For our purposes, let's imagine that the article is running in, to name a magazine not entirely at random, *Scientific American*. Weekly and daily publications for print or digital media have similar systems, but they aren't likely to be as elaborate.

Proposal

When a writer sends in a query letter, offering to write a particular story, the letter will be read first by an editor who makes assignments for that part of the magazine or is responsible for that area of expertise—whether it's meant for the news briefs in the magazine's opening section, for instance, or is a proposed feature story about a mission to Mars. These editors, who make assignments, edit the story once it comes in, and then coordinate with the other departments of the magazine until it's published are known as *assigning editors* or *line editors.* They are usually at the mid- or senior level, so they have at least several years' worth of experience in evaluating ideas, along with the sort of general background that comes from tracking various science and technology fields over time. The masthead may give the titles of such editors as associate or senior editors.

If you're a staff writer at the magazine, the assigning editor is the one who shepherds your story idea, too—but you may not need to write up as elaborate a proposal as a freelance writer would. In either case, staff writer or freelance, if the story is promising, the assigning editor works with the writer to make the query as good as it can be, asking for missing information, and otherwise helping shape the proposal, so that it has a better a chance of being accepted.

Once the query looks good enough, the assigning editor brings it to the attention of one or more *top editors,* whose titles on the masthead include the managing editor, executive editor, and editor-in-chief—or, in the case of *Scientific American,* to a board of editors.

By the way, if you're making your first query to a particular publication, call up to find the name of the assigning editor to whom you should address your query letter. But don't try to make a pitch over the phone. Most editors, who may receive numerous proposals each week, are not going to sit still for a phone pitch unless they already have a good working relationship with the writer. Plus, as you might expect, editors will want to see a sample of unedited writing—that is, the query letter itself—as one of the items we use to help decide whether it's worth the risk to try a new writer.

Assignment

If a proposal is accepted, the assigning editor writes an *assignment letter.* When I write such a letter, I spell out in detail what I expect, including the gist of the story, its due date, and length. I list the requirements for submission of art or illustration materials, and sometimes I even name certain sources who should be interviewed. A word to the wise: If you're working with an editor for the first time and you do *not* receive an assignment letter, do yourself a favor and write one and send it in ("As we discussed on the phone, I will . . ."). When assignments go wrong, it's often because one person has one thing in mind and the other is thinking along different lines. If you're uncertain, it never hurts to ask and to put the agreed-upon result in writing.

Along with the assignment, the editor also mails the writer a contract. At *Scientific American* we retain legal counsel when necessary, as do most magazines, but editors know a thing or two about contracts, too. We have, over the years, gained enough knowledge of contract issues to answer most writers' questions about the rights involved, and to negotiate those rights if necessary.

First Draft

When the manuscript arrives, the editor's "assigning" tasks are finished, and now the line editing begins. The editor gives the article a critical read, and other editors—such as the top editors who originally approved the proposal—will read it and comment too. They want to know the basics: Does the article have a good *billboard*? Also known as the *nut* paragraph (or *nut graph*, in industry lingo), the billboard is the all-important description of what readers will get out of the story if they stick with it. I personally think the billboard is even more important for a science story than for most other types of articles. When I was a general-assignment newspaper reporter years ago, I had the luxury of knowing that my readers would immediately grasp the significance of, say, an art center opening or an arsonist's torching a building. You can't make such assumptions about science topics when you're writing for a general audience.

That's part of the invigorating challenge—and frequent frustration—of writing about science. This stage of editing is when the crucial questions get asked, and answered. Is the angle working? Is there some sort of overarching logical framework, or story line? Beyond a good *lead*, or introduction, does the story have anecdotes and other intriguing material that will serve to pull the reader's eyes through the piece? Is any of the phrasing awkward? Is the tone appropriate for the publication? Does the narrative have good transitions, from section to section, paragraph to paragraph, sentence to sentence? Are there definitions and appropriate analogies, to make sure that the reader doesn't get left behind—or worse, get frustrated enough to give up and turn the page?

Here, no matter how long we've been in the business, science editors must play the part of neophyte readers, looking for any broken links in the chain of information or the flow of the narrative.

Prescription

Frankly, my goal as an editor is to touch as little of the writing as possible—otherwise, what's the point of hiring the writer, who has a unique voice and expertise? But it's rare that a manuscript needs only a few minor adjustments. More often, especially when the article is written by someone new to the business or at least new to the publication, the article will need a revision—at times a significant one. In that case, the editor may draft what is known as a *prescription letter.* This is a note that, like the original assignment letter, aims to tell the writer how to meet the publication's needs. Because the editor now has a manuscript in hand, the prescription letter is likely to be a lot more specific, calling certain sections to the writer's attention. Accompanying the letter will be a marked-up copy of the text, which will include any questions, suggestions for additions or deletions, and ideas for reorganization. The prescription letter is generally written and signed by a single individual—usually the editor who made the original assignment—but it reflects the thinking of a group of top editors as well. At *Scientific American*, we may identify which editor has which question; at other publications, the prescription letter is presented in the voice of a collective "we."

Line Edit

When the next draft comes in, the line editor (and colleagues) will review it again. If we are able to accept the article for publication, the line editor will put into motion the paperwork for payment. (This person usually also codes and submits to accounting any travel and expense receipts turned in by the writer.) Now the editor will complete the line edits—by going through the text line by line to see that all is in order. When that task is completed, the article heads over to the copy desk.

Production

Now the story enters the production machinery of the publication. The story's pictures and words will briefly take separate tracks, before coming together in their final incarnations.

If I'm the story's editor, I serve as the focal point in coordinating all of these efforts, acting as the advocate for both the article and the writer throughout production. I start by meeting with the art director, who will be responsible for the visual presentation of the article. With suggestions from the author, the team will put together a plan for the layout. The editor will usually write the *display type*—the headlines, photo captions, and other elements that draw readers into the story. Writers may (should!) also contribute ideas, though the editor has the final word.

A few words about science illustrations are in order here. In science publications, the art development can seem as intensive as the reporting and writing. If the goal is, for instance, a depiction of how the brain processes visual inputs received from the eye, the editor may sketch a sort of storyboard—a series of pictures and captions that match the writer's vision of the information to be conveyed. Like the story itself, each illustration has to have a main point it is trying to get across. The pictures and captions explain what happens first, second, and so on. Increasingly, science illustrators specialize in certain topic areas, so they may also do research or tap resources to help them transform the editor's or writer's initial sketches into a handsome and informative rendering. Adding to the illustrations, the photo editor may hire a photographer to do a shoot of a scientist on site, search through the portfolios of photographers or the stock images to find just the right ones, or perhaps hunt through micrographs or other handouts from a source's laboratory.

Meanwhile, the story lands at the copy desk. The copy editors will check the text for grammar, spelling, punctuation, and style. Where the line editor is charged with the overall shape, tone, and logic of the story, the copy editor reads it phrase for phrase, even word for word. I suppose the difference between line and copy editing is akin to that between inspecting a feather with your unaided eyes, to see how its colored sections generally work in the overall form, and then peering at that same feather under a microscope to see the actual latching mechanisms.

At many magazines, fact checkers also go over the article closely, meticulously checking each fact, using interview transcripts, reference works, papers, textbooks, and phone calls. This sort of fact checking is probably more common at monthly magazines than it is at publications that have greater frequency. In the case of a breaking story that's going into tomorrow's morning edition—or, for that matter, this afternoon's Web update—the reporter has to be primarily responsible for the facts. That should always be the case, actually. When an error makes it to print, you'd be surprised at how frequently the mistake was typed in at the outset and then somehow was never caught along the way. It's important for each of us to feel personally responsible for the accuracy of the facts we share with our readers, for all of the obvious reasons.

The art and copy soon come together. Nowadays, that happens in some electronic format, which can show the writer, editor, and artist exactly how the printed page will look. The editor fits the copy to the layout, proofreads it, and inserts changes from the fact checkers and the writer, as needed. At the end, copy editors check the changes. The typeset article may go through several versions, or passes, until it is considered ready for printing, and the writer may be asked to sign off on each one—it's a good idea to offer if you're not immediately invited. The printer will later provide a final check in the form of proofs, which today are in full color but which are often still called by their old name: *bluelines*, or *blues* (they looked like a reverse blueprint). Then it's time to roll the presses—literally.

Your real-life experiences with editors won't be exactly as I've described here, of course. Like science itself, the creative process is rarely that neat. For one thing, all these faceless editors will have names, talents, and egos of their own. They may seem brilliant or boneheaded, beneficial—or the bane of your existence. But it's worth any uncertainties or headaches to be part of an enterprise that educates the citizenry in a way that makes them scientifically savvy.

Part Three

Varying Your Writing Style

■ ■ ■ ■ ■

Now that you've read about the basics of how to compose your stories and target your market, you're ready for the next step: thinking about the many writing styles available to you. Different styles will require different talents, and will stretch you in idiosyncratic ways. How much of your own point of view do you want to reveal to your readers? How aggressively do you want to dig behind the press releases and journal articles? How much of your artistry do you want to devote to explanations of the science, and how much to storytelling?

In the next six chapters, you'll find tips on how to write in a variety of styles: daily deadline writing, for which Gareth Cook (chapter 17) advises you to triage your interview list, by calling first the people who will be hardest to reach, leaving for later the ones with cell phones or in an earlier time zone; expository writing, for which George Johnson (chapter 20) suggests you think about your subject like a black box that needs to be unwrapped, one layer at a time; and "gee whiz science writing"—an evocative term used by Rob Kunzig, who tells you how to find, and exploit, "the little nuggets of joy and delight" that made so many of us fall in love with science in the first place (chapter 19). You'll learn from Jamie Shreeve (chapter 21) how the best narrative writers entice their readers "down the rabbit hole" with irresistible opening lines, and from Rob Kanigel (chapter 22) about the "writerly excesses" and "noisy, intrusive yelps of your imagination" that characterize the best science essays. (By the way, Rob's essay about essays is itself one of the best science essays you're likely to read this year.) And Antonio Regalado (chapter 18) will offer his hard-won insights into investigative reporting, complete with guidance about how you can uncover fraud or incompetence not just by combing through documents but, sometimes, simply by asking a source, "How do you know that?"

All of these approaches depend on a kind of suspense, a forward trajectory that keeps the reader paying attention and longing for more. This momentum is most evident in narrative writing, the kind with a beginning, middle, and end. But it exists in the best science writing whatever the style. "Stay tuned," George Johnson says we signal to the reader in our expository pieces. "For now, you will just have to trust me." No matter how complicated the subject matter, this is the implicit pact between reader and writer: "You will just have to trust me." Trust me to take you through it step by step, to show you why it's important to pay attention, to make you *want* to pay attention, to offer active verbs and helpful metaphors and whatever else it takes to get the point across.

Of the various writing styles available to the science writer, each one generates a unique set of reader expectations. If it's an essay, the reader expects a certain amount of attitude; if it's a narrative, the reader expects some evocative details and a fair amount of suspense. If it's a piece of investigative reporting, the reader is ready to be surprised and maybe outraged; if it's a piece of "gee whiz" science writing, the reader is ready to be amazed. But underlying all of these expectations, there's that "trust me" that George talks about, too. This trust we're asking for from our readers demands our most devoted attention. They're counting on us to make our stuff worth reading.

The thing that unites all these approaches is the emphasis on good, clear, clean, original prose. No matter what form you choose, your work won't be of much use if it's not well written. Even under the pressure of a daily deadline, there's time, as Gareth points out, for a beautifully wrought phrase. Don't get too writerly (unless you're writing essays; then a certain propensity to, as Rob Kanigel puts it, "suffocate with language" may be forgiven), but think always about finding the perfect word, *le mot juste*, to get your point across. Look for the way to say something in a completely unexpected way, like the sentence Jamie cites as the great opening to a great piece of narrative writing: "They're worried about the dead man's health."

George Johnson summarizes the task of the good science writer as "explaining the strange in terms of the familiar." I love that description of our work. I love, too, the way Rob Kunzig describes the holy grail of science writing, the subject that many of us spend our careers looking for: one that comes complete with "history, poetry . . . a fascinating central character"—and, by the way, "some nifty science."

One thing that our authors haven't stated specifically is nonetheless important to keep in mind: When you're making choices about the various approaches to science writing, one way to find your best fit is to read, read, read. If a narrative approach appeals to you, read examples not only of great narrative nonfiction but also novels, short stories, even plays. This will give you an ear for the particular language and pace of good storytelling. If you want to flex your muscles by writing science essays, steep yourself not only in those "best essays" collections Rob Kanigel talks about, but also in essays of all kinds: in *The New Yorker*, in memoirs, on the op-ed pages of your newspaper. For investigative writing, go back and reread books like *All the President's Men*—the subject doesn't need to be science-related for a book to have plenty to teach. And if you're looking for inspiring examples of the step-by-step unfolding of a complicated idea—the underpinning of both expository science writing and the completely besotted "gee whiz" approach—consider taking your cues from sources as varied as cookbooks and home repair guides.

By their very variety, and their sheer beauty, the chapters in part III are testimony to how many different ways there are to be a science writer. You can see, even more clearly perhaps than in other parts of the book, what it means to have the "style and voice" you read about way back in part I. The best science writers are brilliant conversationalists, bringing to their work personality and flair. That's what we have in the next six chapters: fascinating conversations (necessarily one-sided, but feel free to talk back anyway) with some of the country's most accomplished science writers, showing you by example how they've learned to do it so well.

ROBIN MARANTZ HENIG

17

Deadline Writing

GARETH COOK

Gareth Cook, a science reporter at the *Boston Globe*, was awarded the 2005 Pulitzer Prize in Explanatory Reporting for his coverage of stem cells. He graduated from Brown University in 1991 with two bachelor's degrees in international relations and mathematical physics. Before joining the *Globe* in 1999, he worked at a number of publications: *Foreign Policy, U.S. News & World Report, Washington Monthly,* and the *Boston Phoenix*. He lives in Jamaica Plain, Massachusetts, with his wife, Amanda, and their son, Aidan.

The moment I walked into the newsroom, I could tell that something was wrong. A group of editors were huddled around the city desk, talking. The televisions were on. People didn't just look tense; they looked genuinely worried. As I walked over to my desk, I saw the image of a burning building. It was the World Trade Center. I was standing there when the second tower fell. I had the same thought that I'm sure a lot of people had: How could this be happening? But I'm also a newspaper reporter, and I realized that there was a science story to be done: Why did the towers fall? Six or seven hours later, I needed to have a finished story that answered that question.

It is hard enough to successfully translate the arcane jargon of science into a story for the general reader. A ticking clock makes it that much more difficult—the words "exciting" and "terrifying" come to mind. For a science reporter, this type of breaking news situation doesn't happen very often. One of the great surprises when I moved to science writing a few years ago was that many of the news stories that appear in daily papers were not, in fact, written on deadline. I used to be in awe that someone had the ability to boil down some complex journal article on human origins or supernovas, reach all the important people, and write a clear, elegant article in a day. Many of the big journals, of course, operate on an embargo system, in which reporters are given advance copies and allowed to report ahead of time on the understanding that they won't publish a story until the journal appears in print.

But there are still times when science news must be delivered on a daily deadline, either because news breaks or because you have a scoop you don't want to lose. In these cases, I think that everyone who does this for a living develops his own set of tools for coping. Success requires a ruthless attention to where you are in the process, where you are in the day, and what you still need. The great enemy in deadline reporting, especially when the material itself is difficult, is panic. Panic means you can't think well. Panic means you can't be creative. A plan helps keep panic at bay.

Over time these deadline tools have become a part of how I think about stories, even when I'm not on deadline. They help me stay focused, even if I have a week, a month, or a year to deliver.

So on the morning of September 11th, I sat at my desk and tried to clear my head. I turned to my trusted tools for deadline reporting. Perhaps some of them may help you.

First, Get Stupid

Before you jump on the phone and into the mad world of reporting and writing as quickly as you can, it is a good idea to take a minute to think about the story in its simplest terms. Why is anyone going to be interested in this, and what are they going to want to know? What pieces of information are absolutely essential to the success of this story? Whom do I absolutely have to talk to? Doing this first is a good way to get moving in the right direction, but it also serves as a kind of rudder as the day progresses. Sometimes you will feel like you are making good progress, finding interesting material, but then you'll get to the end of the day and realize that you have not answered a basic question that readers (and editors) are bound to want answered. This is not a happy place to be.

It is a good idea to come back to these basic questions later in the day. Explaining the story to someone (editor, friend, or significant other) is also a good way to spot holes.

For the story I wrote on September 11th, I knew that people would want to know how the towers fell. So they would want to know what held them up, and how the attack was able to demolish them. Ideally, I would talk to engineers who could answer these questions.

Triage Reporting

From the first minutes of the story, and as the day progresses, I keep a list of calls I need to make, in "triage" order. If someone is going to be hard to reach, for whatever reason, they go high up on the list. If I know I can get them later—they have a cell phone, they are on the West Coast, and so on—then they go lower down.

Often, several of the first few calls I make are not interviews but are to get the ball rolling on people I want, or information I want, that I know will take some time. Keeping this list of calls updated as I go forces me to always keep in mind what I have and what I need, and to have a plan that gets me to the end of the day with a story.

When the space shuttle *Columbia* exploded, some of my first calls were long shots—an acquaintance who might know somebody who used to be high up in NASA, for example. (In this case, it didn't help on the story, but it came in handy later on.)

The Usual Suspects

There are some questions that are good to use in interviews almost no matter what the topic. By the time you are through with these, you are bound to have thought of other questions.

So, filed under "What to ask when you have zero seconds to prepare":

What is new about this?
What is not new?
What is the significance of this, and why?
Who will disagree with this?
What is the evidence this is based on?
Who funded this research?
What will be done next?
Who else should I talk to?
What is your connection to this, and why did you get interested?
How can I reach you later, including in the evening?

Another thing to remember is to resist the urge to jump off the phone. People often say the best things at the end of a conversation. I can't tell you how many times someone has said, at the end of a conversation, "Well, you know"—followed by vital information. Always ask if there is anything else you should know.

Have a Brilliant Idea

Well, maybe not brilliant, but give yourself a little space to do the unexpected. Don't just put your head down. No matter how short the time, there is almost always time to try at least one long-shot reporting gambit (the blank you fill in after "Wouldn't it be great if I got ________?"), and there is almost always time to try something interesting with the writing.

It is, in fact, crucial to give yourself a little time on the writing. Remember that your challenge is to communicate as much information as possible, and bad writing gets in the way of that. For example, I remember I was once doing a daily on a space probe that scientists crashed into the asteroid Eros (the Greek god of love) around Valentine's Day. In a walk to the drinking fountain, I remember thinking it would be fun to write the news as a tragic love story.

The lead became: "Nearly 200 million lonely miles from Earth, and only two days short of Valentine's Day, one of NASA's prized satellites has committed suicide in the name of love."

Clarity

The best news writing is above all clear, and that should be the ultimate goal. That is especially true with science, which is inherently difficult to understand and write about. Make sure that you understand the basic science behind the story, and that you have thought of a way to explain it that will be both clear and accurate. On deadline, I find that I have a kind of running conversation with sources through the day, in which I explain my understanding of some difficult point, and ask them to suggest a better way.

At the same time, you want to try to identify early on what the difficult scientific concepts are going to be and ask yourself: Does the reader really need to know this? If so, can I really explain it so that everyone will get it?

During the course of the day that September 11th, I was able to talk with several engineers. They told me that many of their colleagues were also wondering how the towers could have fallen. They explained the rough outline of what by now has become familiar: that the fire after the impact eventually melted away the structural steel that held the towers up. When one floor eventually gave way, it slammed into the floor below, creating a domino effect they called "progressive collapse."

B-Matter

Another way to ensure clarity is to begin the writing well before the reporting is done. Often there is material—such as history or other kinds of background—that you know will have to be in the story but that is not part of the news. This we call "B-matter," as opposed to the "A-matter" that is the lead of the story.

To take an extreme example, imagine you are assigned to write a news story about a presidential address, scheduled for 8 p.m., about a major change in the government's policy on stem cells. The lead of this story will, of course, be whatever the president says and its implications. But you also know that there are a lot of other things you can write ahead of time: the recent history of the policy, the arguments for and against changing it, why the change is happening now, and so forth. When it comes time to actually write the story, you may need to rewrite some of this; but boy, will you be glad you took a first swing at 3 p.m. instead of 9:30 p.m. when your editor is yelling at you.

Even in more routine situations, there is often an opportunity to get a little writing done ahead of time, while you are waiting for someone to get back to you. Writing is a pretty stark way to learn what you still don't know. It also gives you a chance to come back and improve what you've done after some more reporting (and a mental break).

There is a danger in overdoing this. You don't want to spend too much time writing, because it takes away from reporting time, and you can also end up with more than you will actually need for the story (read: "You wasted time").

Context

In addition to knowing the news, readers need to know why it matters, and why it matters now. Make sure that you have a very good, very clear, answer to those questions and that it appears high up in the story. Often this is what separates the professional version of a story from the amateur version.

Another challenge of deadline writing is writing to length. Editors do not like to contend with a story that is longer than it is supposed to be. Usually, you and an editor will agree how long the story should be during the course of the day. When I write, I often check the length, with the goal in mind. Although getting the length right may sound like another daunting requirement, it can actually save you work. Often I will get to the halfway point in the writing and realize that I will not be able to even talk about several topics.

On big stories such as the one on September 11th, there are a large number of reporters and a smaller number of stories. Typically, one reporter will take the lead on a particular story, and other reporters then send that reporter "files"—sections of story, describing what they have found, which can be dropped into the story that one reporter writes. This happens because the logic of reporting does not always follow the logic of stories that readers will want.

For example, I was concentrating on engineers for my story on the towers, but I also received a file from a reporter who was reporting on the reaction among architects and what the buildings meant, symbolically. This made my story richer and more interesting.

At the same time, reporters putting together broader stories on what happened needed to know from me roughly what I had learned about why the towers fell, because the main stories on this disaster needed to mention, more briefly than I would explain, how the buildings could collapse. Editors referee this process, but like in any team sport, there is no substitute for direct communication with the other reporters, so everyone knows what everyone else is doing.

Accuracy

Terrifying but true: You can't make a mistake, and there are thousands of ways to make one in any article. After I turn in a story to my editor, I print it out and go to a quiet place if at all possible. Then I use a pen to underline every fact I see, adopting the most paranoid mood I can muster.

As I double-check the facts, I mark them off. Discuss any technical explanation with someone who knows exactly how the process you are explaining works, to make sure you have not made some subtle, unwarranted leap. It is also good to walk outside for a minute if you can; this always clears my head and makes me remember something that I wanted to double-check.

Ending Well

Some conventional wisdom holds that the ending of a news story is not that important. This is a vestige of the days when news stories were (literally) cut from the bottom. Obviously, the important things have to go high in a story, but that doesn't mean the ending needs to be a dud.

Try to end the story in the way that will give readers a sense of completion. (This doesn't have to be a quote, either.)

I like to think of structuring a news story like hosting a party. When your guests arrive at the door, welcome them and tell them what is going on, but don't overload them with information. As they settle in, make sure that they get around to talk to the people they should meet. Look like you are having fun, even if you are stressed out. When it's time for guests to leave, say goodbye and maybe even give them a parting gift to remember the occasion by. Who knows, they might just want to come to your next party.

18

Investigative Reporting

ANTONIO REGALADO

Antonio Regalado is a science reporter at the *Wall Street Journal*. He graduated from Yale University in 1991 with a degree in physics and received a master's degree in journalism in 1995 from New York University's Science and Environmental Reporting Program. Antonio's work at the *Journal* was part of a winning Pulitzer Prize submission in the Breaking News category for coverage of the terrorist attacks of September 11, 2001.

When a biotechnology executive whose company I had often written about published a memoir, I got a chance to learn how he saw journalists—in particular, me.

The executive, Michael West of Advanced Cell Technology Inc., in Worcester, Massachusetts, blamed me for some disastrous publicity that had befallen his small cloning company. On page 193 of his book, *The Immortal Cell* (2003), he let me have it: "Antonio Regalado is more of a *detective* than a reporter" (emphasis added).

I think Dr. West was honestly surprised by the lengths to which I had gone to find out about his company's research. ACT was at that time pioneering a controversial technology called "therapeutic cloning." I had gone to the patent office and delved through voluminous files. I had called just about everyone who'd ever worked with the company. I'd asked impertinent questions. I wouldn't take no for an answer.

Although Dr. West's book portrays me as a somewhat dastardly fellow, being called a "detective" is one of the biggest compliments I've ever been paid. What's more, I learned from his comment that the approach to reporting I had taken was very different from that of other science journalists he'd dealt with.

The fact is most science journalists are concerned with *explaining* science to a general audience. Reporters take difficult material and present it in a way that lay readers can understand. With so much of modern life based on science, explaining it clearly is probably our community's most important objective.

But sometimes we science reporters can get a little complacent. We can be too trusting of scientists' good intentions, and we forget to be skeptical. Too often, we allow *Science*, *Nature*, and the *Journal of the American Medical Association* to spoon-feed us the news each week.

Seth Shulman, a reporter who has covered toxic waste and government censorship of science, told me that his definition of an investigative project is "a story that doesn't want to be told." That's why leaked reports or confidential memos so often play a role in investigative reports. Sometimes a physical paper trail is the only way to find out what people were really thinking.

To be a science "detective" requires a more critical view of things. I tend to assume, for instance, that researchers aren't telling me the whole story. I always wonder about hidden motives. And as far as the scientific data goes, I believe that's fair game for tough questions, too. Usually, the facts are benign and the motives innocuous, and the science is okay. But not always. More often than you might think, a little digging will turn up evidence of dodgy ethics, commercial entanglements, or simply bad research.

ACT's research raised red flags in all these areas. Dr. West saw his tiny biotechnology company's project—whose aim was to clone human embryos for stem cell research—as a big medical breakthrough. Others saw it as a step in the direction of fetus farms and carbon-copy children.

The truth is, part of me suspected that the company was secretly interested in farming cloned fetuses for their organs. Was I nuts? That same year, Robin Cook published *Shock*, a creepy medical thriller about a secret baby cloning project at a small company outside Boston. It was clearly inspired by ACT—but it was fiction, not fact.

My quest for the unvarnished truth ultimately took me to the cavernous reading room of the U.S. Patent and Trademark Office in Crystal City, Virginia. As I sat poring over the company's patent applications, suddenly . . . there it was. A written description of how to harvest transplantable tissues from "embryos, fetuses, or offspring, including human."

It wasn't quite a smoking gun. Patents are always written to take into account every conceivable application of a technology, even unlikely ones. Yet I had confirmed my most bizarre suspicion—the scientists involved had at least considered cloning fetuses and babies to harvest their organs.

There was no baby factory to write about, but I did end up using what I learned. A few days after my trip to Crystal City, the *Journal* published an investigative article describing how the Patent Office was struggling to keep ACT and several others from winning legal rights over cloned human embryos. An August 20, 2001, my colleague Meera Louis and I co-authored a story entitled "Ethical Concerns Block Widespread Patenting of Embryonic Advances" that began like this:

> As the science of cloning and embryonic stem cells advances at a breathtaking pace, universities and companies are seeking sweeping patent claims over the new technologies. In the U.S., patent applications in these two areas have jumped 300 percent in just the last year.
>
> But ethical prohibitions embodied in patent law in the U.S. and Europe are preventing scientists from securing patents on some pioneering biological inventions.
>
> At issue in the U.S. is the 13th Amendment to the Constitution, which abolished slavery. Patent documents show that the legal prohibition against owning humans has complicated efforts by Geron Corp. and closely held Advanced Cell Technology Inc. to patent medical uses of human cloning technology.

While a reporter's goal is often to discover what's happening behind the scenes, some of the best investigative stories in science have asked the question: What if the experts are just plain wrong? These aren't your typical investigative stories about "waste," "misdeeds," or "bad guys." In these cases, the journalist is instead going after flawed scientific thinking, theories, or results.

Freelance reporter Gary Taubes has carved out a specialty uncovering and challenging such unfounded beliefs. In two lengthy articles published in *Science* magazine about the health effects of salt and of dietary fat, Gary showed how opinion—rather than facts—was actually behind some widely accepted public health advice.

"The (Political) Science of Salt," published in *Science* in 1998, began this way:

> In an era when dietary advice is dispensed freely by virtually everyone from public health officials to personal trainers, well-meaning relatives, and strangers on check-out lines, one recommendation has rung through three decades with the

> indisputable force of gospel: Eat less salt and you will lower your blood pressure and live a longer, healthier life.

The next 9,000 words of the article were dedicated to showing how that prevailing view on salt was, in fact, not based on hard data at all. The dangers of salt had become the conventional wisdom through force of repetition, not on the strength of the science. In the words of one researcher quoted in the article, the government's anti-salt campaign went "way beyond the scientific facts."

A key factor in such reporting is the recognition that there's often more to science than just science. While theories should be based on data, in practice science can be heavily influenced by factors ranging from money to researchers' egos. In the case of salt, Gary argued the problem was the pressure to turn complex and conflicting scientific data into simple public health sound bites.

Gary gave me this definition of investigative reporting on science: "It's not about uncovering the conspiracy or following the money. It's about do the data really mean what scientists think they mean?"

That kind of reporting isn't easy. Gary interviewed 80 experts and spent a year reporting the salt story. But anyone can implement the basic approach by asking scientists simple questions like "How do you know that?" Pretty soon, you'll start to see where the data are incomplete or inconclusive, and what kind of assumptions often lie behind scientific statements.

The *Science* magazine story on salt, and a follow-up story about dietary fat, both won the prestigious National Association of Science Writers Science-in-Society Award. Those stories and other winners from radio, TV, and newspaper outlets can be read online at www.nasw.org. I recommend them as a starting point for anyone interested in doing in-depth reporting on medical or scientific research.

■ ■ ■

Another rich area for investigative science reporting is any case in which human volunteers agree to take part in scientific experiments, such as tests of new medical treatments.

The rules that govern human-subjects research today trace back to a trial of Nazi doctors that took place in Nuremberg, Germany, after the end of World War II. Some of the doctors had conducted experiments on people in concentration camps. The so-called Nuremberg Code developed during the trial says the quest for knowledge should never harm unwilling and uninformed participants.

It's a delicate balancing act. And these days, big-time financial interests all too often threaten to skew researchers' ethics and good sense. That was what *Washington Post* reporters Deborah Nelson and Rick Weiss found in 1999 when they began investigating the death of 18-year-old Jesse Gelsinger, a volunteer in a gene therapy trial at the University of Pennsylvania. In their first story, published in September 1999, the lead researcher described the death as an unfortunate accident:

> This was a tragic unexpected event," said James M. Wilson, director of the university's Institute for Human Gene Therapy. "I hope in a month we'll have looked at every angle so we can share with whomever is interested in listening what we have learned from this.

It was just so much spin. Digging through journal articles, patient consent forms, and conference abstracts, the *Post* reporters discovered the study had serious flaws. The university had failed to alert the Food and Drug Administration that they had already witnessed serious side effects in monkeys and in two earlier patients. And Dr. Wilson had financial ties to a biotechnology company pursuing gene therapy that might have benefited indirectly had the study worked. In other words, the young patient who died had never been given all the information—a basic requirement for informed consent.

The payoff of the *Post*'s reporting came over the ensuing months. As they pursued the story, Deborah and Rick learned of other unreported deaths at other universities. A federal investigation ensued, studies were halted, and the National Institutes of Health created new rules to make research safer for patients.

■ ■ ■

A key lesson for the *Post* reporters was that scientists had compromised their ethics and good judgment, possibly because of the fame and financial gain riding on a successful outcome. Particularly in biology, financial ties with biotechnology firms are now common. At stake is nothing less than scientists' most highly prized asset: their objectivity.

At a minimum, a reporter should always ask a scientist if he or she has financial interests in the outcome of ongoing research. Many leading biologists are founders of biotechnology firms, and some own stock. Others may serve on so-called scientific advisory boards, which may pay them $20,000 or more a year to give occasional advice and lend their name to a company's efforts.

Most big universities operate "technology transfer" offices whose job is to patent scientists' discoveries and sell them to companies. (The scientists usually get a share of the income.) Try finding out what invention is making the most money for your alma mater. To contact the technology transfer at your school, look it up on the Website of the Association of University Technology Managers at www.AUTM.net.

Sources and Resources

People are the most important sources for almost any story, but finding documents that can help tell a story is a crucial facet of most investigative reporting. Following are three sources of documents useful in investigative science reporting:

Patents and Patent Applications

In 1998, I wrote a story called "The Troubled Hunt for the Ultimate Cell" for *Technology Review* magazine. The story was about the quest to isolate human embryonic stem cells, and in the article I identified a half-dozen teams who were in the race.

At the time (six months before scientists announced they'd found the cells), there was basically zero information on the Web or in the scientific literature about these fascinating cells. And many scientists were working under conditions of near-secrecy. So how did I find them?

It turned out that most had been quietly, but busily, filing patent applications.

The basic notion of the patent system is to encourage innovation by giving inventors a 20-year monopoly over their ideas. But there's a trade-off: To win a patent, an inventor must disclose his or her idea in great detail in a patent application. Those applications, as well as approved patents, are now easily searchable in online databases maintained by the U.S. Patent and Trademark Office (www.uspto.gov) and by the European Union (www.espacenet.com).

Science reporters rarely make use of this giant repository of technical information. One reason is that it's pretty dense stuff. But using the patent databases may let you get a peek at research projects not yet disclosed elsewhere. That's what I did to get a jump on the big stem cell story. I remember one researcher I called, Roger Pedersen of the University of California at San Francisco, at first refused to admit he was even working on stem cells. Wasn't he surprised when I started reading from his own patent application!

Freedom of Information Act

The federal Freedom of Information Act (FOIA) and similar state laws are important tools in the science reporter's kit. For research funded by a federal agency, there are at least some documents obtainable via FOIA that you can't get otherwise.

I've used FOIA to get handwriting samples of a suspect in the 2001 anthrax attacks and to learn how many laboratories are working with the SARS virus. Not every FOIA request will become a story, but it's important to try FOIA in order to learn how the process works.

Recently, I had a suspicion that Roger Pedersen—the same scientist whom I'd previously butted heads with over stem cells—was trying to clone human embryos. As usual, Dr. Pedersen wouldn't talk, and the UCSF public affairs office flat-out denied that anyone at the school was cloning embryos.

I didn't believe them. My *Journal* colleague David Hamilton came up with the idea of using California's state FOIA law, known as the California Public Records Act. Using this tool, we could arguably have gotten all Dr. Pedersen's lab notebooks, maybe even personal e-mails and expense receipts. But with FOIA requests, it's important to target your request as narrowly as possible—that increases the chance you'll get your documents.

We ended up asking for any patient consent forms related to egg donors (eggs are needed for cloning), as well as any documents mentioning human eggs generated by the school's ethics review board. UCSF stalled for 10 months—and *Wall Street Journal* lawyers sent threatening letters—but finally we got the documents. As I suspected, Dr. Pedersen's research group had gotten permission to use hundreds of eggs from the UCSF fertility clinic in cloning experiments!

Two useful resources in how to file FOIA requests are George Washington University's guide to FOIA (www.gwu.edu/~nsarchiv/nsa/foia_user_guide.html) and "How to Use the Federal FOI Act" by the Reporters Committee for Freedom of the Press (www.rcfp.org/foiact/).

Food and Drug Administration

The FDA, which approves and licenses drugs in the United States, is another vital resource for documents. On its website, the agency posts warning letters to drug makers who've failed inspections, lists doctors who've broken its rules, and makes available transcripts of meetings of its outside scientific advisors. Such transcripts can make fun reading—the pros and cons of new treatments are often discussed in the frankest terms.

After a pharmaceutical company wins approval for a new drug, the FDA also makes available the company's filing to the agency. Lawyers and competing companies regularly obtain these extensive files, although few journalists are aware they exist.

For a reporter ready to delve deeply into a topic, FDA documents can be a valuable source of unadorned data. Science writer Stephen S. Hall told me they proved crucial to a cover story he wrote on the allergy drug Claritin for *The New York Times Magazine*. Steve, who suffers from allergies, got interested after his doctor mentioned that he thought the multimillion dollar blockbuster didn't seem to work for many people.

When Steve looked into the question, the FDA documents told the inside story: In its effort to come up with a "nondrowsy" allergy medication, Claritin's maker had marketed the drug at doses so low that it was barely effective against allergies.

The FDA's entire catalogue of approved drugs, and associated regulatory documents, recently became available online. Find it at www.accessdata.fda.gov/scripts/cder/drugsatfda.

■ ■ ■

Investigative reporting is worth doing. It can change how government or industry works, help protect people from harm, and improve our society in important ways. And for many journalists, their investigative reports are the jewels of their careers. They think of it as the best work they have done, and so do their editors.

19

Gee Whiz Science Writing

ROBERT KUNZIG

After a youth spent largely in Europe, Robert Kunzig studied history of science at Harvard, mostly for pleasure. To the extent that he thought about the future, he thought he might be a foreign correspondent. But his academic background made him more attractive to editors as a science writer. A piece for *Newsday* on proton decay led to a job at *Scientific American,* which led to a series of jobs and 14 years at *Discover,* for which Rob is still a contributing editor. Along the way he won several awards, including, twice, the AAAS award for magazine writing, and the Aventis Prize for his book *Mapping the Deep* in 2000. In 1996 Rob moved with his family to Dijon, France. He ended up being a foreign correspondent after all—a foreign science correspondent.

A couple of years ago I learned something: I learned that black holes spin. And as they spin, they drag the fabric of space-time around with them, whirling it like a tornado. "Where have you been?" you ask. "That's a direct consequence of general relativity! Lense and Thirring predicted that more than 80 years ago." It had escaped my notice. It made my day when I (sort of) understood it. I wanted to tell someone—and by a wonderful stroke of luck, I'm paid to do just that.

Days like that are why I'm a science writer—a "gee whiz" science writer, if you like. A lot of my peers these days consider the gee whiz approach outdated, naive, even a little lap-doggish; investigative reporting is in. "Isn't the real story the process of how science and medicine work?" Shannon Brownlee said recently, upon receiving a well-deserved prize for her critical reporting on medicine. "I'm talking about the power structure. I'm talking about influence. I'm talking about money."

I'm not much interested in those things. I agree they're often important—more important, no doubt, in breast cancer than in black hole research, more important the more applied and less basic the research gets. One of the real stories about medical research may well be how it is sometimes corrupted by conflicts of interest. Power, influence, and money are constants in human affairs, like sex and violence; and sometimes a science writer is forced to write about them, just as a baseball writer may be forced with heavy heart to write about contract negotiations or a doping scandal.

Yet just as the "real story" about baseball remains the game itself, the "real story" about science, to me, is what makes it *different* from other human affairs, not the same. I'm talking about ideas. I'm talking about experiments. I'm talking about truth, and beauty, too. Most of all, I'm talking about the little nuggets of joy and delight that draw all of us, scientists and science writers alike, to this business, when with our outsized IQs we could be somewhere else pursuing larger slices of power, influence, and money. The bits of new knowledge, the elegant chains of reasoning, the ingenious designs that would make you say "gee whiz," if anyone still said that. Or "that's the damnedest thing I ever heard," as Dennis Flanagan, my old mentor at *Scientific American*, used to say.

The ever-changing story of what we know about the world and how we come to know it—that's the science story I'm interested in. It's wide-open territory. "Science is what scientists do," was how Flanagan once tried to fob off a philosophically minded Dutch reporter, who had asked him for a definition—but in Dutch it sounded funnier, and it stayed on his bulletin board: *Wetenschap is wat Wetenschappers doen.* I've spent 20 years walking into scientists' labs or calling scientists on the phone and asking them to tell me what they're doing. I expect I'll be doing it for another 20.

Should you be interested in following this approach, I've found certain attitudes to be helpful.

Don't Be Afraid

Scientists can be intimidating; they know so much about such complicated things. As an editor at *Scientific American*, though, I soon discovered that one thing most *Wetenschappers* don't *doen* very well is write. It was sometimes almost sweet, how incompetent they were—how unable to offer a clear, logical account of their work that would be understandable and interesting to an intelligent layperson. And yet writing, I think we can all agree, is one of the highest manifestations of human intelligence. The lesson: The scientist knows more than you (about his subject, anyway), but he can't do what you do. Each of you is doing something important. The two of you need each other. There's no need for either of you to get shirty.

Keep Your Brain Open

There is a strong pressure in our business, as in science itself, to specialize. It's easier because it cuts down on the per-article learning curve (which can be very important, I'm now discovering, when you're a freelance). It may give you a better shot at scoops, because it's hard to cultivate contacts in lots of disciplines. And you may be a person who is passionate about one field. But I've always resisted specialization, to the point even of straddling the great physics–biology divide. I've written features for *Discover* on particle physics, astronomy, space exploration, oceanography, neuroscience, genetics, biomechanics, archeology, paleoanthropology, and currency reform, among others. Starting afresh in a new field is fun; it helps keep you excited about what you do, and thus able to share that excitement with your readers. Whereas specialization can lead to boredom and jadedness, both of which are fatal.

One of my professional weaknesses is I'm not a font of story ideas; I have some good ones from time to time, but I also get a lot from editors. (I realize this isn't a luxury you have when you're starting out.) That way, I find myself getting interested in subjects that might not have seemed up my alley.

One of my best pieces came about that way. A few years ago a colleague at *Discover*, Sarah Richardson, suggested I write about an Icelandic biotechnology company founded by a geneticist named Kári Stefansson, who wanted to do genetic analyses on the entire Icelandic population. The story seemed littered with medical applications, ethical issues, even potential conflicts of interest; in other words, it did not sound promising to me at all. But I did it anyway. I discovered that besides all those things it had history, poetry, some nifty science, and a fascinating central character. It stretched me tremendously.

More recently, Sarah wrote to ask whether I'd like to write about some mysterious conical gold hats from the Bronze Age in Germany. Absolutely, I said.

Remember That Ideas Are as Interesting as People

Often more. If you don't believe that, why would you be a science writer? Yet we all tend to write a lot of profiles. Sometimes it's because the particular scientist really is fascinating; sometimes a profile just offers a ready-made narrative thread. As I've gotten older, though, I've gotten less comfortable with the whole idea of swooping into someone's life for a day or two and then turning it into a narrative, as if I actually knew the person. Plus, I hate trying to describe what people look like.

It's a challenge, but you can construct narratives that are based almost entirely on ideas and how they developed. Lately I've been choosing to keep the ideas narrowly focused, because that allows you to go deeper, and to delay as long as possible that frustrating moment when you throw up your hands and admit you can explain no more. A few years ago I read a *Nature* News and Views by Frank Wilczek, the physicist, in which he described some recent accelerator experiments that tended to confirm a theory he had helped develop called quantum chromodynamics. I got an extremely satisfying feature out of that—out of describing how the inside of a proton could be made up mostly of gluons and emptiness rather than solid quarks, and how it is that physicists think they know that.

Similarly, the cover story I wrote for *Discover* on spinning black holes was essentially a deconstruction of a single paper in the *Monthly Notices of the Royal Astronomical Society*—actually of a single figure. (*MNRAS* is not on my bedside table; it was a press release that tipped me off.) That spiky little graph purported to show the first evidence that black holes whirl space-time, and in the process act as giant electromagnetic generators. Gee whiz! I thought. How can they possibly say that? I shudder to admit this to an audience of journalists, but I did the whole story by phone and e-mail, without leaving my desk. I still don't know what those guys look like.

Make It Beautiful

Most scientists are motivated at least in part by beauty—you should be, too. Read your prose aloud. Quietly in your office with the door closed—the point is not to alarm the people around you, but for you to hear what your words sound like. They should sound beautiful. At least one passage in each piece should sound really beautiful. Contrary to what some other writing coaches might tell you, that's not the passage you should delete before you even turn it in—on the theory that if you like it so much, it must be purple. That's the passage you should stand next to with a baseball bat when the editor comes tromping through in muddy boots.

One of the vilest enemies of beauty is routine. Most of us, most of the time, write articles that sound like most of the rest of us. Fight this. Editors will badger you for "you-are-there" leads followed by "nut grafs" or "billboards." Subvert their intentions (respectfully). There are other possible leads, other possible approaches—including ones that no one has used before.

In 1994, for example, John Seabrook wrote a piece for *The New Yorker* called "E-mail From Bill." I didn't then and don't now give a hoot about Bill Gates, and I don't remember what Seabrook said about him. What I remember is how he said it: The whole article was structured around his e-mail exchanges with Gates. It was fresh and new to me, and I wished I'd thought of it.

There is beauty to be found not just in individual passages but in the whole structure and rhythm of a piece. When I outline a piece I think first of the transitions between sections, which ideally will correspond to visual breaks on the printed page. Those breaks are your biggest punctuation marks, and like commas and periods, they create rhythm. After that I look for cross-links between the sections—I don't think of the article as a series of paragraphs joined by one long thread, but as a network or fabric. If I know, for instance, that I'm going to need a particular source late in the piece, I try to avoid waiting until then to introduce him, only to hustle him right back offstage; I look for a connection that will allow me to introduce him earlier. It sounds trivial, but I'm convinced that sort of thing creates resonances that make the whole piece feel more satisfying.

When I am feeling most acutely the poverty of my gifts, I sometimes pick a volume of poetry off the shelf, where it has been gathering dust. I open it to some beautiful passage, and I read it aloud, and I tell myself, in my small and modest way, I can do something like that. Usually it doesn't help, of course. But it's a nice way to waste a few minutes.

Have Fun

You might even try to be funny. I don't mean go for Dave Barry belly laughs. That probably isn't compatible with the demands of scientific expository writing. But any time you can lighten things up a bit, you help readers feel more comfortable with material that is inevitably hard work. When I was in Iceland to write about the genetics work that Kári Stefansson was doing, I found time to read the Sagas. I noticed that Stefansson, an ambitious and competitive scientist who also wrote poetry, bore a certain passing resemblance to Egil, one of the most famous heroes of the Sagas—at least enough resemblance for me to slip the following passage into my *Discover* article:

> Egil composed his first sassy poems at age 3, and throughout life he retained the habit of answering straightforward questions with spontaneous eruptions of verse. He also showed his warrior promise early. In *Egil's Saga*, in a chapter called Egil at the Ball-Game, the 6-year-old hero learns to cope with defeat at sport: "Grim had just caught the ball and was racing

> along with the other boys after him. Egil ran up to him and drove the ax into his head right through to the brain."

I would have found a way to use the Egil story no matter what; after all, I'm someone who thinks a high point of cinematic history was when Danny DeVito struck Billy Crystal in the head with a frying pan in *Throw Momma From the Train*. But as it happened, the Egil scene set up a passage later in the piece, in which Stefansson dodged a question about ethical implications of his work by quoting Auden at me, and then proposed (in jest) to throw me in a nearby lake.

How wonderfully strange scientists sometimes are! When I was working on a piece on fetal brain development, Pasko Rakic, a neuroscientist from Yugoslavia and Yale, told me how he had learned his trade from Paul Yakovlev, of Russia and Harvard. Yakovlev had established a library of thousands of human brains, which he embedded in celloidin and then sliced thin like salami:

> Years later, when Yakovlev died, leaving instructions for his own brain to be sectioned and added to his collection, Rakic went to pay homage. "I looked in auditory cortex, and I see pyramidal cells," he remembers. "And I said, 'I talked to those cells.'"

When they say something like that, find a way to use it, even if it's a detour.

Be Skeptical

Of course. If you weren't skeptical by nature, why would you be a science writer? Be skeptical of what scientists tell you the way they are skeptical of one another. Be critical when their statements and actions require it. By all means, follow their money when your nose tells you that's appropriate.

But always remember that 95 percent of scientists—and since I'm making the number up, I might as well say 99 percent—are honest, well-meaning, not seriously conflicted people. They're ordinary people (most of them) who are doing something extraordinary. They're out there beavering away on the edge of knowledge, and they have extreme difficulty explaining what they are up to. Every week they send back cries for help in *Nature, Science,* and *PNAS*. I mean, Gee whiz! It sometimes amazes me that the world should be set up that way, just for me. But I guess there's room in it for you, too.

20

Explanatory Writing

GEORGE JOHNSON

George Johnson writes about science for the *New York Times from Santa Fe, New Mexico, and is a winner of the AAAS Science Journalism Award. Two of his books, Strange Beauty: Murray Gell-Mann and the Revolution in 20th-Century Physics* (1999) and *Fire in the Mind: Science, Faith, and the Search for Order* (1995), were finalists for the Aventis and Rhone Poulenc Science Book prizes. His latest book is *Miss Leavitt's Stars: The Untold Story of the Woman Who Discovered How to Measure the Universe* (2005). He is co-director of the Santa Fe Science Writing Workshop and a former Alicia Patterson fellow. A graduate of the University of New Mexico and American University, he started out covering the police beat for the *Albuquerque Journal.* He can be reached on the Web at talaya.net.

I remember with some precision when I began believing that there is nothing so complex that a reasonably intelligent person cannot comprehend it. It was a summer day, when I was 15 or 16, and my best friend, Ron Light, and I decided that we wanted to understand how a guitar amplifier works. We both played in a mediocre 1960s-era garage band. While Ron went on to become a fairly accomplished guitarist, I was slowly learning that any talent I had didn't lie within the realm of music. Already the aspiring little scientist, I was able to learn enough of the logic of basic harmony theory to execute the mindlessly simple algorithms called bass riffs, and if pressed I could even fire off a bass solo, the dread of concertgoers everywhere. But my approach to the performance was purely intellectual. I didn't have rhythm, or maybe soul.

Poring over the symbols on the circuit diagram of Ron's Fender Deluxe Reverb amplifier seemed infinitely more interesting than trying to read music. I wanted to know what that impressively convoluted blueprint really meant, how electricity flowing through the labyrinth of wires and components could cause the tiny vibration of a guitar string to be multiplied so many times that it rocked the walls of the living room, inciting the neighbors to call the police.

This was still the era of the vacuum tube, before those glowing glass envelopes were replaced by coldly efficient transistors and microchips. Electronics was pretty simple to understand. I had already learned some basics from *The Boys' Second Book of Radio and Electronics* and the guide for the Boy Scout electricity merit badge (the colorful embroidered patch was decorated with a human fist clutching zigzag lightning bolts). In a typical circuit, there were resistors that, true to their calling, resisted electricity, pinching the flow of electrons. There were capacitors, also aptly named, that stored electrical charges. There were tightly wound coils of copper wire called inductors that would hold energy in the form of electromagnetic fields. Finally, there were the vacuum tubes themselves, mysterious pockets of illuminated nothingness inside of which the actual amplification took place.

At first the detail and complexity of the schematic, showing how all these parts fit together inside the Fender's vinyl-covered wooden cabinet, were overwhelming. I could feel my mind start to shut. But with the help of some slightly more advanced books from the Albuquerque Public Library, I realized that I was taking the wrong approach. The trick was to break down the diagram into pieces, master each one, and then put them back together again.

Before long I could place my finger on the diagram and follow the path of the vibrating electrical signal—a replica of the sound of the twanging guitar or the thumping bass—as it traveled through the maze of squiggly lines. Each of the mysterious vacuum tubes, I came to see, was nothing more than a lever. The minuscule fluctuating voltage emerging from the guitar was fed to the first tube, where it was used to operate a gate that controlled a second, much bigger voltage. What resulted was a larger copy of the original signal. This was sent on to the next tube and leveraged again. Step by step, the undulating swings were transformed into ones wide enough to move the cone of the loudspeaker, which would ripple the air and shake your eardrums and stimulate the auditory nerve—a kind of neural guitar pickup that turned the vibrations back into electricity again, input for the brain. By the time I was in college, I could zero in on a malfunctioning circuit and repair it. I could add tubes to the output stage of a lowly Deluxe Reverb, turning it into a more powerful and expensive Super

Reverb. I was amazed that I could get so far with just the broad outlines of understanding.

Whether I was trying to comprehend the workings of a television, a digital computer, or the molecular circuitry inside a cell, the technique was the same: Draw a line around a small portion of the mechanism and treat everything inside as a black box. Color it solid black, if you'd like, for now you will ignore whatever is inside. You can take it on faith that, given a certain input, the box produces a certain output. Later on, if you like, you can pry off the lid and zoom in closer for a more detailed view. Or you can pan outward, lumping the pieces into bigger and cruder chunks. Most people look at a whole TV as one big black box that takes signals from the air and magically turns them into sound and pictures. Any device, no matter how complex, can be understood on many different levels of abstraction.

I didn't appreciate back then that I was already approaching the world like a science writer.

Suppose you want to describe how a brain cell, or neuron, works. In an early chapter of a book about memory, I gave myself the luxury of two fat paragraphs:

> Each neuron receives electrical impulses through a treelike structure called a dendrite, whose thousands of tiny branches funnel signals into the cell. In computer jargon, the dendrite is the neuron's input device. While some of the arriving signals stimulate the neuron, others inhibit it. If the pluses exceed the minuses, the neuron fires, sending its own pulse down a stalk called an axon. The axon is the output channel. It feeds, through junctions called synapses, into the dendrites of other cells . . .

That seemed like just enough to create a mental picture of the basic mechanism without scaring off too many people. Then, to advance the narrative as quickly as possible, I engaged in some hand waving, glossing over a century of research with a sentence I hoped would entice readers with the promise of what lay ahead: "The resulting circuitry is complex beyond imagination. A single neuron can receive signals from thousands of other neurons; its axon can branch repeatedly, sending signals to thousands more. . . ."

Then an initial evocation of what synapses are:

> While information is carried inside a neuron by electrical pulses, once the signal reaches the end of the axon it must be ferried across the synaptic gap by chemicals called neuro-

> transmitters. On the other side of the synapse, the dendrite contains structures called receptors that recognize these transmitting molecules. If enough are registered, then the second cell fires. . . .

As the book unfolded, I would unwrap more boxes, revealing microscopic ion channels opening and closing, triggering physical changes in the cells. I'd describe the molecular cascades that strengthen the synapses, linking the neurons into the circuitry that encodes new memories.

But for now I was content to drive home the point with a simple coda: "A neuron can be thought of as a cell whose specialty is to convert chemical signals to electrical signals, then back to chemical signals again."

That was the shard I wanted to lodge in the reader's mind.

■ ■ ■

More often you must evoke a phenomenon more compactly. In a piece for *Time* magazine, I barely had the leisure to remove the outer wrapping:

> Scientists have long believed that constructing memories is like playing with neurological Tinkertoys. Exposed to a barrage of sensations from the outside world, we snap together brain cells to form new circuitry-patterns of electrical connections that stand for images, smells, touches, and sounds.

With a considerable expanse to cover in 2,000 words or less—the newest theories of how experience leaves its mark on the brain—I had to leave the neuron itself inside its box. It was enough to think of it as a unit to be combined with other units to form the neurological maps called memories.

Whether you are writing a newspaper story, a book, or something in between, the procedure is the same. You start with all the wrappings on. With a few verbal brush strokes you rough out a mental picture, activating a few neurons in the reader's brain: Superstring theory, you begin (this was for the *New York Times*), is "a kind of mathematical music played by an orchestra of tiny vibrating strings. Each note in this cosmic symphony represents one of the many different kinds of particles that make up matter and energy."

A bit later, you peel back another layer: "To give the strings enough wiggle room to carry out their virtuoso performance, theorists have had to supplement the familiar three dimensions of space with six more—curled up so tiny that they could be explored only with an absurdly powerful particle accelerator the size of an entire galaxy."

You're on your way. Never mind, for now, why it takes vast energies to study extremely small things. Don't explain too much too soon: "It's a fact of life on the subatomic realm that smaller and smaller distances take higher and higher energies to probe."

Hint to your reader: Stay tuned. For now you will just have to trust me.

I never did get around in that story to a good, crisp explanation of the energy–size connection. For the material that was to follow, the hand waving seemed enough. But the idea can be evoked with a metaphor. An ordinary microscope cannot resolve things much smaller than a single cell—light waves are too big and clumsy. Focusing more finely requires the shorter wavelengths of an X-ray microscope. X-rays, of course, are more penetrating than is visible light (smaller wavelength = higher frequency = more energy), so you can extrapolate: The smaller the object to be illuminated, the higher the frequency and the more powerful the beam.

Not every physicist is going to like that (though in fact it was a physicist, Maria Spiropulu at Fermilab, who suggested the analogy to me). All kinds of quantum mechanical subtleties have been swept under the rhetorical rug. Sometimes you just have to settle for a good approximation. We are interested outsiders writing for other interested outsiders using metaphor instead of mathematics. It is nice work if you can get it, explaining the strange in terms of the familiar.

The mathematician John McCarthy has a saying he likes to append to his Internet postings: "He who refuses to do arithmetic is doomed to talk nonsense." Sometimes even a science writer must include some very simple math in a story. Presented the right way, the numbers come alive and take on the character of metaphor.

When I wrote *A Shortcut Through Time* (2003), I was faced with evoking the potential power of an invisibly small experimental device called a quantum computer. One consisting of a string of just 64 atoms would, in theory, carry out 18 quintillion calculations at the same time. For a "conventional" supercomputer like one recently built at Los Alamos National Laboratory to do that, I wrote, it would need millions of trillions of processors:

> And so, all things being equal, it would occupy 750 trillion acres—roughly a trillion square miles. It wouldn't fit on the planet. The surface of the Earth is just 200 million square miles, so a supercomputer as powerful as the invisible 64-atom quantum calculator would fill the surfaces of 5,000 Earths, assuming you could figure out a way to operate equipment on ocean-floating platforms.

In my footnotes (called the "Fine Print," a kind of running gloss on the nature and limits of science writing), I showed how I arrived at this figure and poked a little fun at the attempt:

> [The Los Alamos computer has] about 12,000 processors in a space of half an acre. So say that the full one-acre floor would hold 24,000 processors, and roughly speaking, the whole computer would do that many calculations at the same time. So to do 18 quintillion calculations the area would expand by a factor of 18×10^{18} divided by 24×10^{3}, which comes out to about 750 trillion acres. A square mile is 640 acres, so we end up with more than a trillion square miles, 5,000 times the size of the surface of the Earth. Now actually, a single processor (though basically a serial calculator) can perform more than one operation during each machine cycle, so maybe the imaginary machine would occupy merely a thousand Earths. And perhaps before long the processors will be 10 times faster. So that brings us down to a hundred Earths. That's how it goes with these back-of-the-envelope calculations. The point of all the arithmetic is just to say that it would be very big indeed.

A science writer is ultimately an illusionist. The conjuring is in the service of a noble cause: getting as close as linguistically possible to scientific truth.

Adapted from *A Shortcut Through Time: The Path to the Quantum Computer*, by George Johnson (New York: Alfred A. Knopf, 2003).

21

Narrative Writing

JAMIE SHREEVE

Jamie Shreeve's most recent book is *The Genome War: How Craig Venter Tried to Capture the Code of Life and Save the World* (2004). A 1979 graduate of the Iowa Writers Workshop, Jamie contributed fiction to various literary magazines before turning to science writing. From 1983 to 1985, he was public information director at the Marine Biological Laboratory in Woods Hole, Massachusetts, where he founded the MBL Science Writing Fellowship Program. His previous books include *The Neandertal Enigma: Solving the Mystery of Modern Human Origins* (1995), and with Donald Johanson, *Lucy's Child: The Discovery of a Human Ancestor* (1989). A contributor for *The Atlantic Monthly, National Geographic, The New York Times Magazine, Wired,* and other publications, he lives in South Orange, New Jersey.

"The universe is made of stories, not of atoms," wrote poet Muriel Rukeyser. While some physicists may not agree, the power of narrative to grip a listener's attention is certainly ubiquitous in human society, and its roots run deep into prehistory—perhaps as far back as language itself. There is even some evidence that the brain is hard-wired to remember information better if it is transmitted in narrative form. As a science writer, you should exploit this propensity whenever the opportunity arises.

At its simplest, a narrative is a sequence of events with a beginning, a middle, and an end. In this sense, narrative is endemic to science itself—the creation of the universe, the life history of a butterfly, and the action of an antibody on a pathogen are all essentially narrative events. Broadly speaking, a scientific paper is also a narrative, with a hypothesis (the beginning) determining methods (a middle) that lead to results and conclusions (the end). You may, in fact, have occasion to structure a story by tracing the unfolding of some natural process, or by following the protocol of some specific experiment.

In most cases, however, good narrative requires something more than just a beginning, middle, and end: notably characters, and some scenic context for them to interact in, and some implicit tension—a conflict between two scientific teams after the same prize, for instance, or the heartbeat-by-heartbeat progress of an experimental surgical procedure with a life at stake. Developing character and building tension require space, which is why most narrative science writing is to be found in magazine articles, newspaper feature stories, and books. But a journalist on a shorter deadline can still incorporate narrative elements into his or her story, especially in the lead as a hook, or sprinkled into the body of the story to leaven the density of the science under discussion.

Whatever the dimensions of your story, tempt readers with a compelling beginning, or they will never make it to your middle or end. Your opening is the rabbit hole through which the reader falls out of reality and into your narrative world. You want to create a disequilibrium in the mind, an itch that can only be resolved by burrowing deeper into the story. Here is the opening to a very well known account of a major discovery:

> I have never seen Francis Crick in a modest mood. Perhaps in other company he is that way, but I have never had reason so to judge him. It has nothing to do with his present fame. Already he is much talked about, usually with reverence, and someday he may be considered in the category of Rutherford or Bohr. But this was not true when, in the fall of 1951, I came to the Cavendish Laboratory of Cambridge University to join a small group of physicists and chemists working on the three-dimensional structures of proteins. At the time he was thirty-five, yet almost totally unknown . . .

This is, of course, the beginning of James Watson's inimitable *The Double Helix.* Watson's rabbit hole is the hinted-at character of Francis Crick, especially his immutable immodesty juxtaposed against the dramatic change in his reputation caused by the story we are evidently about to hear. We have very little notion yet who Crick is, what he did that catapulted him from obscurity to being talked about with reverence, or what his relationship is to the narrator. But we sure do want to find out. In the chapters that follow, we will learn a great deal about DNA and the theory and experimentation that went into unraveling its structure. But it is the interaction of characters, not molecules, that keeps us turning the pages. "Chiefly [the tale of DNA] was a matter of five people," Watson tells us in his preface. "Maurice Wilkins, Rosalind Franklin, Linus Pauling, Francis Crick, and me." Make characters the matter of your narrative, too, and let the science spill from their relations.

Of course, most narrative science writers do not have the advantage of having discovered the structure of DNA or otherwise lived the events in their accounts. The first two chapters in Dava Sobel's collection, *The Best American Science Writer 2004*, illustrate two approaches to creating powerful leads. Sometimes the pursuit of a story can itself provide a robust structure for a first-person narrative. Jennifer Kahn uses this device in her story "Stripped for Parts" in *Wired* magazine, which opens like this:

> The television in the dead man's room stays on all night. Right now the program is *Shipmates*, a reality-dating drama that's barely audible over the hiss of the ventilator. It's 4 a.m., and I've been here for six hours, sitting in the corner while three nurses fuss intermittently over a set of intravenous drips. They're worried about the dead man's health.

What compels the reader here is the deadpan discordance between the everydayness of the hospital room scene, and the fact that the patient is a corpse. Note the deft use of detail: In good narrative, it is not only okay to take the space necessary to tell us *which* TV show happens to be running, what other noises there are in the room, where the narrator is sitting and for how long, and so forth—it is essential. Note, too, that Kahn is in no hurry to reveal why the nurses are fussing over a dead guy—an example of good narrative pacing, which is intrinsic to storytelling. In fact, we don't find out that the story is about organ donors until the third paragraph. By then, we are deep down into this rabbit hole.

In most cases, you will be better off using the third person in your story and not cluttering up the pages with an intrusive "I." Here is another powerfully paced opening, this one in third person, from Atul Gawande's story "Desperate Measures" in *The New Yorker*:

> On November 28, 1942, an errant match set alight the paper fronds of a fake electric-lit palm tree in a corner of the Cocoanut Grove night club near Boston's theater district and started one of the worst fires in American history. The flames caught onto the fabric decorating the ceiling, and then swept everywhere, engulfing the place within minutes. The club was jammed with almost a thousand revelers that night. Its

> few exit doors were either locked or blocked, and hundreds of people were trapped inside. Rescue workers had to break through walls to get to them. Those with any signs of life were sent primarily to two hospitals—Massachusetts General Hospital and Boston City Hospital. At Boston City Hospital, doctors and nurses gave the patients the standard treatment for their burns. At MGH, however, an iconoclastic surgeon named Oliver Cope decided to try an experiment on the victims. Francis Daniels Moore, then a fourth-year surgical resident, was one of only two doctors working on the emergency ward when the victims came in. The experience, and the experiment, changed him. And because they did, modern medicine would never be the same.

Gawande's cool, unadorned style conveys the horror and intensity of the fire that night much better than if he were to start throwing around words like "horror" and "intensity." In fact, the only time he heats up his prose is in the last two lines, where we discover that this story is not really about the Cocoanut Grove fire, but about an individual whose experience changed modern medicine. That's well-planned pacing.

Here's just the first sentence of another opening of a science narrative: "I met my first Neandertal in a cafe in Paris, just across the street from the Jussieu Metro Stop." I confess I am proud of that sentence, which begins my third book, *The Neandertal Enigma.* On the other hand, it took me two years to write it. My point is not that I am a slow writer (which is also true), but that your narrative will only be as good as the research you put into it. You have to familiarize yourself not just with the science, but with the personalities of the scientists and other characters who literally embody your story. Even if you are simply employing narrative in a lead or as a loose framework for what is essentially an expository piece, you still need to know enough about what your characters look like and think like to impart a sense of their individuality. If you are writing a full-blown narrative, you need to become intimate with their backgrounds, their hopes and fears, strengths and weaknesses, foibles, tics, grievances, grooming habits, and manners of speech, not to mention their method of hailing a taxi, removing a hair from their coffee cup, and unwrapping a candy bar. (This is, of course, in addition to the research required to write confidently about the science they are pursuing.) You may find that after two weeks, two months, or two years of research, 95 percent of the details and observations you have jotted down will not be used. Never mind: The act of collecting the information will give authority to that other 5 percent that brings your characters to life and keeps the reader turning pages.

Generally speaking, there are two ways to go about getting this information: either by observing events firsthand, or by interviewing participants after the fact. If you are lucky, you can seize on a story that is still unfolding in real time. In this case, try to place yourself as close to the action as possible, for as long as the participants will tolerate your presence. My last book, *The Genome War*, is an account of the race for the human genetic code between the government Human Genome Project and Celera Genomics, a private company. I was able to obtain exclusive access to Celera from the beginning of the project, and more or less had the run of the place for the next two years. If you can secure a similar fly-on-the-wall position without sacrificing your journalistic independence, go for it. But keep a low profile. Scientists are not going to be put at ease by a writer jabbing a tape recorder in their faces and asking questions in the middle of an experiment, especially when you are still an unknown quantity. Establish a rapport with as many of the people as possible—not just the big shots leading the project, but their colleagues, postdocs, grad students, technicians, secretaries, janitors, and security personnel. Your goal is not to become part of the gang. Your goal is to become part of the furniture. The sooner you are trusted, the sooner you can fade into the background. Observe your characters carefully—a tiny habitual gesture might turn into a revealing detail in your text. One of my characters in *The Genome War* had a habit of using an index finger to scratch inside his ear occasionally—especially at meetings when something was being said he might not want to hear. Meanwhile, keep on the hunt for narrative opportunities—they may not be where you thought they were at the start. When I began research on the book, I assumed that the dramatic tension would spring from the race between the government and the private company. While this very public conflict did supply a great deal of the drama, I discovered that tensions within the company itself were just as potent for driving the story forward.

In contrast to the scientists at Celera, the leaders of the Human Genome Project did not allow me to witness their activities firsthand. In such a circumstance, or if the action of your narrative has already run its course before you begin your research, you must rely on interviews of the participants for the details you need to reconstruct the narrative events and character interactions. Of course, you want to pose questions that will elicit information about what happened. But you also want to gather information on where it happened, what precisely was said, what was the tone of voice used to say it, the expression on the speaker's face, and anything else you can think of that will help you reconstruct an accurate depiction of the event. Let's say there was a meeting that proved to be a turning point. Who sat next to whom? What was the shape of the table, the color of the rug, the landscape outside the window? Even if these details turn out to be inessential themselves, asking them may inspire your source to remember others that turn out to be crisp and usable. Try to coax your sources into a narrative frame of mind. Then stop asking questions and let them tell *you* the story. Substantiate these recollections with those of other participants in the action and with written records, such as meeting agenda and transcripts.

Remember, too, that an important element of a narrative may be what's going on inside your characters' minds as the action proceeds. "If you ask a person, 'What were you thinking?'" Richard Preston writes in a note prefacing his bestselling narrative *The Hot Zone* (1994), "you may get an answer that is richer and more revealing of the human condition than any stream of thoughts a novelist could invent." Don't forget to ask this critical question.

When it comes time to begin writing, one advantage of narrative is that the mere chronology of events provides at least the basis of your story structure. If you have done your research well, the most difficult problem may be selecting which scenes to include and which to leave out. What are the threshold moments in the action—meetings where decisions were made, encounters in hallways or on street corners that had repercussions farther up the narrative path? What interactions best reveal the personality and motivation of your characters? Where is the story's natural climax? Can you work back in time from that point to help select the material that best leads up to it? Don't forget that once your readers are down the rabbit hole and into your narrative frame, most of them *want* to be kept on edge until the final resolution of the main story thread. Remember to pace the reader: Dole out clues cunningly; drip-feed developments. Leave the main story line hanging sometimes in suspense at the end of chapters or sections. If you are working on a book or an extended magazine story, there may be side stories you can turn to before picking up the thread in the main tale.

Near the beginning, you may have to bring in background material that will help the reader understand what is happening in the present moment. There will also be times when you have to interrupt the narrative to explain the science involved. Try to find situations where these expository sections can flow from the action. A biographical digression on a scientist that includes some discussion of her earlier research can serve the dual purpose of illuminating her character, and providing the scientific background your reader requires to understand the research in the narrative present moment. Think scenically. Perhaps there is a context within the narrative where you can fold in some science—a character explaining his research to some laypeople at a meeting, for instance, where his actual level of dialogue matched what you need to reach your reader.

In *The Demon in the Freezer* (2002), Preston—a master at writing about morbidity—has many opportunities to describe the devastating effects of smallpox on the human body. But he also needs to tell his readers what is happening in the disease on a biochemical level. Note how in this passage he sets up such a discussion with a riveting narrative scene:

> When the Sisters of Mercy opened the door of [a smallpox patient's] room, a sweet, sickly, cloying odor drifted into the hallway. It was not like anything the medical staff at the hospital had ever encountered before. It was not a smell of decay, for his skin was sealed. The pus within the skin was throwing off gases that diffused out of his body. In those days, it was called the foetor of smallpox. Doctors today call it the odor of a cytokine storm.
>
> Cytokines are messenger molecules that drift in the bloodstream. Cells in the immune system use them to signal to one another while the immune system mounts a response to an attack by an invader. In a cytokine storm, the signaling goes haywire, and the immune system becomes unbalanced and cracks up, like a network going down. The cytokine storm becomes chaotic, and it ends with a collapse of blood pressure, a heart attack, or a breathing arrest, along with a stench coming through the skin, like something nasty inside a paper bag. . . .

Look for creative opportunities like this to sneak the scientific exposition in under the rug—and, of course, keep your writing vivid with strong verbs, good metaphors, and keen detail. Your readers may end up so entranced with your story they don't even notice they're learning at the same time.

22

The Science Essay

ROBERT KANIGEL

Robert Kanigel has written hundreds of magazine articles, essays, and reviews for *The Sciences, The New York Times Magazine, Johns Hopkins Magazine,* and dozens of other publications. He is the author of *Apprentice to Genius* (1986), about mentoring among elite scientists; *The One Best Way* (1997), a biography of Frederick Winslow Taylor, the first efficiency expert; and *The Man Who Knew Infinity* (1991), *a New York Public Library "Book to Remember"* and a finalist for both the National Book Critics Circle Award and the Los Angeles Times Book Prize. He received the Grady–Stack Award for science writing from the American Chemical Association in 1989 and the Author of the Year award from the American Society of Journalists and Authors in 1998. In 1999, Kanigel became professor of science writing at MIT, where he directs its Graduate Program in Science Writing. He is finishing a book about leather and its imitators.

The essay is a genre-buster.

Nonfiction genres—article, book review, memoir, news report—form a kind of taxonomy, like that a biologist imposes on the animal kingdom, or an astronomer on celestial objects. Yet the essay is a genre that subverts the idea of genre. It's not news. It bears a personal stamp, demanding something of the writer's insights, experiences, or idiosyncratic take. But once past these slim criteria, to call it "essay" says precious little about it.

The science essay can be formal, even stately, as in *Science* editor-in-chief Donald Kennedy's long, sustained argument on climate change, originally presented as a lecture. It can be amusing, as in Alan Lightman's reminiscence of how a failed college electronics project made him, a budding physicist, an ex-experimentalist.

It can suffocate with language, as in Richard Selzer's sense-rich explorations of anatomy in "Mortal Lessons: Notes on the Art of Surgery" (1976).

> I sing of skin, layered fine as baklava, whose colors shame the dawn, at once the scabbard upon which is writ our only signature, and the instrument by which we are thrilled, protected, and kept constant in our natural place.

It can deal with life and death, the cosmos and infinity. Or it can be a slight thing, as in an elegy for the slide rule that I wrote around the time the pocket calculator was supplanting it:

> Long nights spent working physics and chemistry problems would reveal each rule's mechanical idiosyncrasies, the points in its travel where the slide slipped smoothly and those where it snagged. No two rules were alike. Borrow a friend's—same brand, same model, perhaps purchased minutes apart at the student bookstore—and you'd feel vaguely ill at ease. It wasn't yours: *The rough spots were different.*

The science essay can be spartan and simple. Or it can delightfully digress, as Stephen Jay Gould's so often did. "To the undiscerning eye," Gould wrote once, barnacles are "as boring as rivets."

> This is largely attributable to the erroneous impression that they don't go anywhere and don't do anything, ever. The truth of the matter is that they don't go anywhere and don't do anything merely sometimes—and that, other times, barnacle life is punctuated with adventurous travel, phantasmagorical transformations, valiant struggles, fateful decisions, and eating.

The essay can be a small gem like J. B. S. Haldane's 1928 classic, "On Being the Right Size." It can be a long, melancholy memoir, like G. H. Hardy's *A Mathematician's Apology.* It can celebrate ambivalence. Or it can bubble over with adamant conviction, like Clifford Stoll's *Silicon Snake Oil.*

The essay is so open, so much *it can,* that it's scarcely limited to essays. Which is a silly enough way of saying that to see it only as the leisured, ruminative sort of thing that winds up in essay collections is to cruelly constrict it. The book review is an essay whose subject is a book. The op-ed is an essay of about 800 words appearing opposite the editorial page in a newspaper. The introduction to a book is an essay framing the history, reportage, or hard science in the book's body. So, at the other end, is its conclusion.

Most books include stretches that, while not essays per se, bear that individual stamp, that relief from the drum roll of fact, date, and quote, that *sound* like essays. Take Richard Preston's 1987 account of the Hale telescope at Palomar Observatory, *First Light*, where he imagines the heavens as a

> palimpsest containing stories written on top of one another going back to the origin of time. A telescope looking outward into lookback time strips layers from the palimpsest; it magnifies and reimages small, faint letters in the underlayers of the manuscript. The sky could also be imagined as a book, bound into chapters that tell a story. As a telescope probes out into the sky, it reads backward through the story, from the last chapters to the first.

This same *tincture of essay* can be found in many a good magazine article, as well.

The essay with science or scientists as its subject, of course, imposes special challenges, the most obvious being the sheer intellectual orneriness of nucleotides, fractals, enzymes, and quarks. What's a poor essayist to do? Face down the scientific content? Slide around it with metaphor and analogy? Ignore the hard stuff altogether? Do you stand at the door to the lab, point to gels, columns, and centrifuges, but stay right where you are? Or march right in and get your hands dirty?

Of course, how deeply to plunge into the science is a conundrum in all science writing, one that doesn't vanish just because you're writing an essay and not a news report. Sometimes—weary realism speaks here—the decision is made for you. Because, as writer, not scientist, you don't know enough, or don't know deeply enough, to have much to say; best to keep your essayist's mouth shut. But maybe sometimes you do have something to say about nanoengineering or Rosalind Franklin that ranges beyond the facts. Say it? Run the risk of sounding foolish? *That* is the essayist's big challenge.

■ ■ ■

These days, as scientific breakthroughs and faux-breakthroughs compete for headlines and scientific superstars tread the halls of university fiefdoms, readers confront quarks and quasars, protons and prions. They stumble over scientific notation and organic nomenclature; are baffled by incomprehensibly large distances and minutely small ones; get lost in the dark borderland between science and pseudoscience; confuse what they know, what they think they know but don't, what they once knew but no longer do, and what they never knew at all. The reader, then, needs help, and the science writer supplies it. But the science essayist is freer to furnish context that rises above the day's run of press releases from medical center and university press offices; to turn to her knowledge of fields distant from science, medicine, and technology; to look to the past and speculate about the future; to exploit the power of language.

And, always, to invoke her own life experience.

An essay about the sense of touch might profit from recollections of a suede jacket made for you and the feeling of the tailor's hands on your body. One about pollution might benefit from your own experience of sludge washing over you at an otherwise pristine beach. All writing relies on both the *out-there* of datum and fact, and the *in-here* of the subjective, the eccentric, and the personal. The essay gives freer rein to the latter.

Of course, there's a time and a place for everything; the art of essay writing relies, in part, on knowing when to write one. An AIDS breakthrough in *Nature*? The reader probably doesn't want to hear much of anything but the facts, thank you. That first day we probably care little about the history of the discovery or the angst of the researcher, much less ours as writer. But later, even two days later, is another story. Maybe a reflection on a kindred discovery 40 years earlier? Or one on the central role of a crucial new lab technique? Each is a potential essay, one that places the sensibilities of the writer in the service of some fresh perspective that goes beyond today's, or even last year's, headlines.

In a biography I wrote of Frederick Winslow Taylor, the efficiency expert and apostle of scientific management, I chose not to slow up the narrative of his life with analysis of its significance. Yet analysis it needed: Taylor's influence has been felt all through the industrialized West, and his ideas of efficiency shape modern life. And so, in an abrupt shift of tone and emphasis late in the book—but while Taylor, in book-time, was still alive—I launched into a "Report From the End of the Century" (inspired by the ending of Margaret Atwood's *The Handmaid's Tale*) that let me grapple with the legacy of Taylorism in something like a formal essay.

To grapple. Writers always grapple with their material. To make sense of it. To pay homage to all its elements. To unearth the story from a mess of dry fact. The essay, as a form, celebrates such grappling. Drawn from the French *essayer*—to try, or attempt—it's as if the word itself honors our struggle.

In my long essay I ranged far from Taylor's life and into scientific management's impact, its diffusion across national borders, its place in the history of technology, its roots in the American Progressive tradition. In a section called "The Fifteen Unnecessary Motions of a Kiss," I looked at the cultural position of the efficiency expert, and asked what was so scientific, anyway, about scientific management. I compared Taylor with Henry Ford, looked to Taylorism's place in Soviet and Nazi totalitarianism.

Too much? That's certainly one pole of what can go wrong in an essay, and it may have gone wrong in this one—that the essayist ranges too widely, leaving a mash of ideas, insights, factoids, and opinions that lack focus, or scarcely belong in the same piece, or are simply more than the reader can digest.

But I think the opposite danger—*too little*—is, if not more common, more insidious. Here, the result is not "wrong," unprofessional, or lacking in craftsmanship. But it lacks ambition. The writer stops short, limits the territory over which he lets his mind roam, resulting in something simplistic, pat, or just uninteresting.

We know the enemies of good science reportage—flat leads, turbid prose, out-and-out inaccuracy. As a science essayist you face all these, but others, as well. Like not letting enough of yourself into the piece. Or imposing too much of yourself and so indulging in empty solipsism. Or coming to premature closure—not essaying, not grappling, but settling on the first easy answers that flit across your brain. Or never quite getting your hands around what you're trying to say. Or deluding yourself into thinking you have more to say than you do. Hmm.

I'm not sure that the essay—that form destroyer, that rule defier—rewards analysis. It's not that it *can't* be analyzed, broken down, and codified, examples of each type studied, plumbed for patterns. In *The Art of the Personal Essay*, Philip Lopate cites the analytic meditation, the diatribe, the mosaic, the memoir; doubtless some scholars do this for a living. But I'm not sure the would-be essayist learns much from such analytical scrutiny. Rather, exploiting the riches of the essay means giving up every last clinging-to-form, means setting out on an adventure to you're-never-quite-sure-where. David Quammen begins his 1992 essay "Vortex" with a near-death experience in the rapids of the Gallatin River in Montana, likens whirlpools to the great spinning brushes of a car wash, invokes laminar flow, shear stress, and Reynolds numbers, draws us down into the anatomy and physiology of heart valves, and tells how Leonardo da Vinci studied the path of grass seeds sprinkled onto the river, thereby depicting

> the patterns of graceful turbulence that water can assume as it flows against an obstacle or over a drop: curls, eddies, foamy pillows, long waves with the crests peeling over, spirals within spirals, all of these shapes layered down upon one another to give a sense of translucent depth.

We relish every free-ranging word and thought.

But what all this damned *freedom* implies is that writing an essay may seem different and new to those raised on the stern discipline of newspaper journalism or even long-deadline magazine writing. The sharp focus and well-defined subject that in almost every other kind of writing are desirable, even demanded, limit the essay's range and ambition. Not that the essay excuses flabbiness; in the end, it demands the same rigor and incessant self-editing of all good writing. But you may not want to reach for those trusty professional tools *too soon.* You may choose to tolerate, for a while, your writerly excesses and keep open, a little longer, to the noisy, intrusive yelps of your imagination.

Part Four

Covering Stories in the Life Sciences

■ ■ ■ ■ ■

In part IV (and part V), you will find science writing at its most specialized—journalists explaining how they cover astronomy, nutrition, mental health, earth sciences, and more, all in deliberately sharpened focus.

It almost belies the fact that most of us begin as generalists, writing about a wide range of scientific discoveries and trends. There's a natural logic to that: A short attention span and a restless intellect seem to come naturally to journalists. We enjoy that ability to shift from field to field and focus to focus.

Specialization—choosing to build expertise in a particular field of science—usually comes later. Many freelances make that choice; it allows them to build on accumulated knowledge, rather than researching a new subject for each assignment. Large newspapers often prefer their science writers to pursue that narrower path as well, sometimes hiring physicians to report on medicine, for instance.

But although this describes my own career trajectory—evolving from all-purpose newspaper science writer to a journalist specializing in the behavioral sciences—I want to begin by emphasizing the value, and even importance, of science writers who do it all.

It was Lee Hotz, of the *Los Angeles Times*, who brought this to my attention at a science meeting several years ago. He described science writers as the last generalists in the community of science. He proposed—and I've come to agree—that a journalistic preference for the panoramic view of a wide-angle lens is completely underestimated.

Unlike scientists themselves, often tucked into the compartments of their subspecialties, science writers are by nature interdisciplinary. We tend to see—and we often like to make—connections between varied fields. That view from above, as it were, allows us to identify trends that link disciplines together, even seeing patterns that scientists might not.

In fact, Emory University primatologist Frans de Waal, who writes for both a popular and a scientific audience, once said that he is often able to see his own science more clearly when he's writing for lay readers; he's less bogged down in trying to anticipate research objections and more able to concentrate where he wants.

So, although we, the editors, have tried in parts IV and V to provide a wide range of essays on specialty reporting and how to do it well, these contributions should be kept in context. We offer them in the hope that they, too, will be useful in an interdisciplinary way.

For a generalist, the specialties covered here may all belong to a single job description. For those who hope to become truly specialized, the issues raised may give a better sense of what a particular line of reporting may entail. For those who already concentrate on a particular field of research, we hope you may yet find helpful ideas in some of the companion fields or find a fresh look at your own line of work.

And although these chapters explore some sophisticated science beats, there's one other point of context that I want to emphasize. You will not find separate chapters, in this part or the next, in the fundamental sciences of chemistry, biology, and physics.

There are very few science writers who have the luxury of only pursuing the mysteries of physics, of only illuminating the sizzle of chemical reactions. Instead, these sciences are woven through everyone's work: We write about the biology of a virus, the chemistry of a toxin, the physics of a weapons system detonating close to home or a supernova exploding far, far away.

Of course, sometimes we are fortunate enough to find a really good story that's pure, basic, elegant science.

The challenge, then, is always to make our readers (and our editors) appreciate why our explanation of, say, the crystalline structure of ice is really the stuff of journalism. Sometimes that's as simple as suggesting a future application—that the strength of a crystal structure may help us engineer stronger buildings. Most scientists can tell you that; they may even write it into a grant application.

For further insight, though, I consulted with Phil Yam, at *Scientific American*. As news editor, Yam frequently is called upon to deal with the challenge of turning a basic research story into a compelling magazine article. His first recommendation was to consider profiling the scientist doing the work—using the theater of his or her story to lend drama to the elements of research.

He went on: "My favorite kind of basic research story, however, is one that overturns what I learned in textbooks, or at least runs counter to conventional wisdom. That organisms could make a 21st amino acid, going beyond the 20 standard ones, or that proteins, and not just DNA and RNA, can act as elements of inheritance, or that adult mammals can grow new neurons, are all textbook-changers."

What he says, of course, could apply also to any of the specialized beats that follow, once again highlighting the interdisciplinary nature of science writing.

We editors have separated these two parts of the book into "stories in the life sciences" and "stories in the physical and environmental sciences." I would like to acknowledge in advance that these are arbitrary divisions. No one really believes that life sciences are an entirely separate entity from, say, environmental research.

Still, the chapters in part IV are clearly related both in science and in substance. Shannon Brownlee's incisive opening chapter on covering medicine sets what I think is just the right tone—a balance of smart, look-beneath-the-surface reporting and awareness of the people affected by medical decisions—which means every one of us. Paul Raeburn's troubling assessment of mental health treatment in chapter 26 pursues similar themes, in an area of science that deserves far more attention than it gets.

And this balance of good science reporting and good analysis continues throughout. Antonio Regalado's essay on covering human genetics (chapter 28) emphasizes beautifully the need to explore this branch of science with a cautious and skeptical eye; Kevin Begos (chapter 27) offers an insightful look at behavioral sciences and their often dark past; Marilyn Chase (chapter 24) provides a thoughtful historical perspective on treatment of infectious disease; Steve Hall (chapter 29) gives us an elegant exploration of the politically fraught topics of cloning stem cell research; and Sally Squires (chapter 25) offers provocative reporting on nutrition issues. In fact, throughout part IV, you will find this group of outstanding science journalists not only writing about the science itself but also underlining the real-life ethical dilemmas that come in tandem with scientific discovery.

There was a time when you wouldn't find such morally complex perspectives in a book about science writing. In the earlier days of science journalism, the emphasis was almost entirely on making sure the science was correctly explained—which is still a worthwhile goal. But as we've learned more about our craft, we've learned that explaining science correctly also means recognizing that it is a human endeavor, a sometimes fallible process, and that good reporting encompasses that part of the story, as well.

Without blowing our journalistic horn too loudly—this recognition of the balance between science and its ethical fallout is an area where science writers can sometimes outdo scientists themselves. We have the advantage of standing back with our wide-angle lens and observing from a distance. It's taught us that, whether we are specialists or generalists, illuminating the research world, from its brightest spots to its darkest corners, is one of the most valuable contributions that we can make.

DEBORAH BLUM

23

Medicine

SHANNON BROWNLEE

Since leaving *U.S. News & World Report* to freelance in 1999, Shannon Brownlee has written about medicine, health care, and biotechnology for such publications as *The Atlantic Monthly*, *Discover*, *The New York Times Magazine*, *The New Republic*, *Time*, and *The Washington Post Magazine*. A winner of the Victor Cohen Prize for Excellence in Medical Science Reporting and the National Association of Science Writers Science-in-Society Award, Shannon is now a senior fellow at the New America Foundation, where she is focusing on the links between the lack of scientific evidence in medicine, the poor quality of U.S. health care, and spiraling costs.

Medical writers have gone through a period of soul searching, a reappraisal of our role as journalists and members of the fourth estate. Are we supposed to simply cover the medical news: the new findings, the "breakthroughs" that appear in medical journals? Or are we also supposed to serve as critics of medicine, uncovering corruption and wrongdoing like our colleagues who cover politics, the military, and business?

When I started in this business in the early 1980s, we medical journalists liked to talk about ourselves as translators. Our job was to sort through the medical journals, decide what was newsworthy, and then put the jargon of science and statistics into language that ordinary readers could understand.

In the intervening years we've done a superb job of translating and conveying information. In fact, we might have done the job too well, because in simply reporting each newsworthy finding in the professional journals, the lay press has helped sell medical products and procedures to a public eager for good news about their health. The upshot is that we've inadvertently helped put a high gloss on medicine, rather than actually keeping the enterprise honest.

As medicine has become increasingly commercial and political, medical writers have increasingly assumed the role of critic and watchdog. We still have to cover the medical news, but we also have to provide the social, political, and scientific context for each new finding. These days, getting a medical story right requires more than simply understanding molecular biology, or clinical trial design, or how to express relative risk versus absolute risk. Getting it right also means understanding the role that industry plays in driving medical science. It means questioning assumptions about how disease works.

Do a Nexis search for the words "C-reactive protein" and "heart disease," for instance, and you will find dozens of stories that say, in effect, that C-reactive protein (CRP) is the latest and greatest new predictor of heart disease. But what you won't easily find in all that ink are questions about whether CRP is any better than current predictors of heart disease, like serum cholesterol levels or stress tests. You will see even fewer stories that ask whether CRP screening tests will help people avoid heart attacks, or will simply lead to more and more patients being treated with cholesterol-lowering statin drugs.

To write one of the more probing, analytical, skeptical articles about a development like CRP, a few things must happen:

1. You have to understand the science. What is CRP, anyway, and why might it be involved in heart disease? At the very least, read the abstract and the conclusions of the study.
2. You have to know something about the background behind the study. Why did someone think CRP was worth investigating in the first place? For that, you'll have to go to researchers.
3. You have to find at least one source willing to cast a skeptical eye over the study and tell you why the author's interpretation of the results might not be the only alternative.
4. You have to ask who is likely to benefit from reporting the results of this study. A good place to start is by looking at who funded it. Was it the NIH, which generally funds research that is aimed at furthering the basic understanding of disease? Or was it a drug company, which may want to use the results as a marketing tool for a test for CRP?

Of these four tasks, the last one will probably feel the most foreign to medical reporters. More than half of the clinical research in this country is funded not by the federal government but by the pharmaceutical or biotech industry. As the proportion of research that is privately funded has steadily risen over the last two decades, so has private industry's control over what medical research gets done, what gets published, and consequently what you tell your readers. Many reporters are unaware, for example, that the pharmaceutical industry regularly withholds data that might have a negative impact on sales. Pharmaceutical marketing departments often help design studies. And while we hate to think that academic scientists might be influenced by money, numerous studies show that scientists conducting industry-funded research are significantly more likely to find benefit from the sponsor's product than those doing research funded by the federal government or other independent sources.

Here's the bottom line for medical reporters. Pharmaceutical companies, and to a lesser degree the biotech industry, regularly use financial relationships with academic researchers, the people upon whom the nation depends for unbiased information, to distort medical science. All too often, drug companies see academic clinicians as potential spokespersons for their products, and treat the medical journals as mere marketing tools. As Dr. Richard Horton, editor of the prestigious British medical journal *The Lancet,* put it in a March 2004 essay in *The New York Review of Books*, "Journals have devolved into information-laundering operations for the pharmaceutical industry."

That means that you have to be aware of whose bread is being buttered when you report the results of a new study. Many scientific-seeming studies are in fact what Dr. Bernard Carroll, a professor emeritus at Duke, has labeled "experimercials." An experimercial exists not to advance science but to increase market share of a drug that has already been approved. Many experimercials do manage to come to statistically significant conclusions, but they fail nonetheless to provide much useful data. A recent study, for instance, enrolled several hundred elderly depressed patients, half of whom were given the antidepressant Zoloft, and the other half of whom received a placebo. The study found that the mood of the group on Zoloft improved ever so slightly more on average than that of the placebo group, and the difference was statistically significant. But if you really read the fine print, you would find that the patients themselves did not, on average, feel any better.

These days journals, particularly those catering to specialists, are filled with experimercials. Who can blame the pharmaceutical manufacturers for wanting to do this? Once published, these studies allow drug makers to tout the conclusions in their marketing to doctors. If the experimercial gets reported uncritically in the media, the company gets "direct-to-consumer" advertising without having to pay for it.

■ ■ ■

Now that we know some of the rules and pitfalls of reporting on the medical literature, let's look at how the press did a superb job of reporting one new study but flunked when it came to putting the study into context. The study was announced at a scientific meeting in March 2004, and stories about it ran on the front page of everything from the *Times-Picayune* to the *New York Times.* Here's the first line from the front-page story that appeared in the *Washington Post*: "High doses of a popular cholesterol-lowering drug can sharply boost protection against having or dying from a heart attack, according to research that many experts said is likely to transform the treatment of the nation's leading killer."

Right off the bat, readers know this is big news. The *Post* reported that the study, a clinical trial testing the effects of high doses of two statin drugs, Lipitor and Prevachol, found that patients who took higher than normal doses of a statin, and consequently achieved "ultra-low" cholesterol levels, were 16 percent less likely to suffer a heart attack or die than were patients who achieved supposedly "normal" cholesterol levels on normal doses. It went on to quote Dr. Christopher P. Cannon, the cardiologist at Brigham and Women's Hospital, in Boston, who led the study, as saying, "What this tells us is that treating cholesterol is very important, not just for high-risk patients but for everyone."

Now, there's a clear take-home message: Achieving ultra-low cholesterol levels with higher doses of statins is probably a good idea for everybody who is at any risk of heart disease.

But that's not what the study actually says, and here is where many reporters misled readers in their headlong rush to write up the good news. The researchers looked at a very specific group of patients: people who already had serious, symptomatic heart disease. It looked only at the sickest patients—none of whom were women.

That means that the higher doses of statin drugs may reduce the chance of heart attack only in men; and among men, only in those who already have serious, symptomatic heart disease. There is no evidence from this study that people who aren't seriously ill will reap the same benefit, if any.

Second, most of the stories offered readers no real sense of what the 16 percent reduction in risk really means. If you go back to the original paper, you'll see that the chances that the group on a standard dose of a statin would suffer a cardiac event during the study period was 26 percent. It was 22 percent for the group on the higher dose of the drug. So the absolute change in risk was only 4 percent.

Why did so many in the medical press fail to make clear the study's shortcomings? Because journalism is set up to trumpet the good news. Neither editors nor the general public are interested in a story that says, in effect, "Small advance, nobody saved." (I'm borrowing this insight from my colleague Joe

Palca of National Public Radio.) That means that all parties, scientists and journalists alike, want to accentuate the positive, or at least the dramatic. A 16 percent reduction in relative risk sounds a lot better than a 4 percent reduction in absolute risk. (For a refresher on these terms, see chapter 3, Understanding and Using Statistics.)

The other reason the press got the story wrong is that the results, as reported, confirm what we think we already know. Why go looking for a source who will be critical of this study, when it's perfectly obvious that if these drugs can help the sickest patients, they can probably do even more good for those who have milder disease?

■ ■ ■

So what's a medical reporter to do? Imagine that a study has just been released by a prominent journal, and you have to write it up for your publication. The odds are good that the study was funded by a drug company, and many of the top researchers who would be the obvious choices for interviews might have conflicts of interest that may skew their opinions—making them less-than-perfect sources. But you need to move fast because your editor is breathing down your neck to get your copy in.

Your first stop should be the Cochrane Collaboration, a nonprofit international consortium of clinicians, researchers, and statisticians who assess evidence for the efficacy and safety of many procedures, tests, and drugs. The Cochrane Collaboration is simply the best source of unbiased, evidence-based information about the current state of knowledge on medical practices ranging from Alzheimer's drugs to knee surgery. You have undoubtedly run across Cochrane analyses even if you didn't know that's where they came from. For instance, a Danish study created a flap three years ago over the effectiveness of mammography. That was a Cochrane study. You can find Cochrane studies at www.cochrane.org. Ask your news organization to subscribe to the Cochrane Library if you don't already have access to it.

Once you've looked at Cochrane's assessment, go to a patient advocacy group for further insight into the context of the new paper. It's hard to think of a disease that doesn't have a patient group, but you should know that not all groups are created equal. These days, you can never be sure which ones are truly grass-roots organizations that are acting as watchdogs on behalf of patients, and which ones are "astroturfs," which are created by industry to help market a disease. "Freedom From Fear," for example, was founded in 2001 by GlaxoSmithKline, the manufacturer of Paxil, and the public relations firm Cohn & Wolfe to market Paxil for treating generalized anxiety disorder. Reporters all over the country received press releases from Freedom From Fear and interviewed patients provided by the group when Glaxo launched its marketing campaign. The net effect was that Glaxo expanded the market for a drug with potentially serious side effects to people who might not really need it.

If you want to avoid astroturfs, first go to the Center for Medical Consumers, at www.medicalconsumers.org. For a strongly antidrug view of any psychiatric medication, go to the Alliance for Human Research Protection, at www.researchprotection.org. If they haven't looked at the subject of your story, they will know where to send you.

Now that you have a sense of what the critics have to say, you can call up and interview the lead author of the paper with a much better sense of what holes may exist in the study and its conclusions.

Now comes the hard part. At some point during your discussion with the scientists, you should ask questions about the relationship between the researchers and the manufacturer of the product. Be forewarned, academics are not accustomed to such questions, and many will bridle at the suggestion that their research might have been biased by who paid for it. Nevertheless, we have to start asking these questions if we have any hope of understanding the context of the study and how we should write about it. Besides, asking tough questions is your job, so hitch up your pants and dive right in with the following:

First, ask it straight out: Where did the funding for the study come from?

Let's say the study was funded by industry. Your next question should be who initiated the study—the researchers or the company? In other words, whose idea was it? If the company initiated the study, how much control over the study design and the data did the company have? Did the researcher you are talking to see all the data? Did he or she analyze it?

Next question: Who actually wrote the journal article, the scientist or the company? If the scientist actually admits the paper was ghostwritten by the manufacturer or a PR firm, it's doubly important that you find sources willing to scrutinize the study's findings with a critical eye.

And finally, does the researcher you are talking to, or any of the paper's co-authors, have any financial ties to the company? Do any of them have a consulting contract, for example, or own stock or stock options? Is the researcher a member of the company's speakers bureau? Some journals, notably the *New England Journal of Medicine* and *JAMA*, require researchers to disclose this information, but not all researchers comply. It's a good idea to ask.

Asking questions like these could give you a very different story, one that will better serve your readers. Health care, like medical journalism, is going through a period of reappraisal, but neither doctors nor Congress can bring about meaningful reform without the help of the press. Our job isn't to trumpet every tiny advance but to shine light into the dark corners where big business and medical science intersect.

■ ■ ■

Four books might help get you thinking about the issues raised in this chapter:

The $800 Million Pill, The Truth Behind the Cost of New Drugs, by Merrill Goozner (Berkeley: University of California Press, 2004).
Better Than Well, American Medicine Meets the American Dream, by Carl Elliott (New York: W.W. Norton & Company, 2003).
Should I Get Tested for Cancer? Maybe Not and Here's Why, by H. Gilbert Welch (Berkeley: University of California Press, 2004).
The Rise and Fall of Modern Medicine, by James Le Fanu (New York: Carroll & Graf, 1999).

24

Infectious Diseases

MARILYN CHASE

Marilyn Chase, a health reporter for the *Wall Street Journal,* graduated with high honors in English from Stanford University and earned a master's degree in journalism from the University of California at Berkeley. She began covering health and medicine full time for the *Journal* in the early 1980s and now focuses on infectious diseases. She is the author of *The Barbary Plague: The Black Death in Victorian San Francisco* (2003).

Every story of an infectious disease outbreak contains many stories, each illuminating a different aspect of this powerful intersection between people and medical science.

There is the patient's story, of a body under assault by microscopic invaders, and a struggle to recover and live.

There is the germ's story, of how a bacterium or a virus spreads from its refuge in nature to invade the bodies of animals or people, traveling through the blood, settling in organs, and wreaking cellular havoc.

There are the laboratory dramas, of masked or space-suited researchers growing and analyzing germs to identify, to characterize their habits, and finally to find the Achilles heel that enables an effective counterattack.

There are the doctors' stories of grasping all available tools to diagnose and treat. When tools are lacking, it's a story about inventing new tools, at first crude and full of side effects, then later more sophisticated tools including targeted drugs and vaccines. There are tales of clinical trials, of researchers who partner with volunteers balancing hope of cure with risk of harm, which also contain undercurrents of ego and altruism.

There are the company stories, of corporate officers who take these discoveries and gear up to test, manufacture, distribute, and—of course—profit from drugs or vaccines. There are the regulators' stories, of the Food and Drug Administration's attempts to balance the urgency of a green light for needed treatments with the mandate to uphold safety and meet the Hippocratic requirement to "First, do no harm." And there are stories of political and financial motives here, as well.

These perspectives—personal and political, social and financial—often clash. Some of the most compelling stories involve such conflicts.

The global fight against AIDS is rife with clashes between scientific and social goals. One recent conflict is centered on the study of an AIDS treatment that might be used to prevent infection. The drug, called tenofovir, is being tested among people who are unable to or who choose not to avoid exposure to the human immunodeficiency virus (HIV) by means of condom use, sexual abstinence, or fidelity to one partner who is uninfected.

Among the risk groups who might need such a drug are prostitutes and sexually active gay men who are unable or unwilling to use other means of protection against HIV. No one knows whether the drug will work. But the possibility of a preventive pill—a concept technically known as "pre-exposure prophylaxis" or PREP threatens to undo years of public health messages aimed at getting people to change their behaviors. One expert source shared the anxiety of many in the field, saying: "We don't want the message to be, 'Take this pill, and then go —— your brains out.'" Recently, PREP trial sites in Asia and Africa were shut down after protests that researchers didn't provide adequate protection or treatment for volunteers.

Denial of outbreaks is a perennial theme in infectious diseases as bureaucrats view an epidemic as a source of stigma, tarnishing political prestige, and eroding trade and tourism dollars. The reluctance of governments to recognize the threat of mad cow disease out of concern for the beef industry risked more suffering and economic losses in the end. The slow response of many countries to the AIDS crisis helped fuel the spread of the pandemic that by 2004 was affecting 40 million people worldwide. The initial reaction of China during the SARS outbreak in the winter of 2002–2003, when the government initially downplayed cases of a strange new pneumonia, impeded swift global control measures. When, in the spring of 2004, SARS resurfaced, it renewed anxiety worldwide.

But no country stands alone in epidemic denial. This misguided impulse has been a recurrent theme ever since the Black Death hit Europe in the mid-14th century. The cycle of denial, delayed response, further spread, and increased suffering has been replayed ever since around the world.

In the United States, politicians have conducted cover-ups of their own. I recently spent several years documenting an outbreak of plague in 1900 San Francisco for a book. Trade ships introduced plague rats to the city, where poor Chinatown residents suffered a deadly outbreak and were made scapegoats for their sufferings. City and state officials blamed the patients for breeding disease, blamed the diagnosing doctors for inventing a crisis, and delayed an effective plague control program for almost a decade. The cover-up was orchestrated by the state governor, the mayor, and the chamber of commerce. The result of official denial: Plague jumped from urban rats to rural squirrels and spread eastward to the Rocky Mountains. While some newspapers colluded with corrupt officials, other reporters fought the politicians' "pact of silence" by printing the truth.

Fast-forward to today: The neglect of the 1900 outbreak still exacts a price. Plague germs allowed to spread a century ago still linger in rural rodents throughout the Southwest, especially in the Four Corners region of New Mexico, Colorado, Arizona, and Utah. Indeed, people contract plague from contact with infected fleas each year. If not diagnosed and treated quickly with antibiotics, it still leads to devastating illness and death.

Unfamiliar with the warnings of history, politicians and industrialists have in recent years downplayed the threat of mad cow disease, SARS, and avian flu. As often happens, vulnerable patients pay the price. Today as in years past, reporters can play a key role in bringing accurate and balanced public health information to light.

How to go about reporting an outbreak? Interview doctors treating the outbreak at its source, but don't limit yourself to official channels or to any one kind of source. It's a good idea to develop layers of sources in infectious diseases, epidemiology, lab sciences, public health, and patient advocacy. It's helpful to develop a well-rounded cadre of experts on any given disease—people you trust as your sage heads to help you unravel what is going on, and to explain its broader significance.

Big public health agencies—such as the National Institutes of Health, the Centers for Disease Control and Prevention, and their counterparts in countries all over the world, plus the World Health Organization in Geneva—are the spokes of a network of reference laboratories that receive patient samples for analysis, confirm diagnoses, declare epidemics underway or over, and coordinate campaigns of eradication and control. But as big institutions, they don't convey the human scale of an epidemic—so you still need local sources on the ground.

After the terror attacks of 9/11, as the specter of bioterrorism arose, I spent several days hanging around a state laboratory in Berkeley, California, profiling microbiologist Jane Wong as she went through the steps of confirming or ruling out a possible germ attack. The story resulted from a routine call I had made to the California Department of Health Services—even before the anthrax letters had come to light. As a medical reporter, I was simply interested in advanced preparations for a hypothetical threat. The lab allowed me to come in and observe up close, in a way that would later have been impossible, once the anthrax panic caused many such labs to lock down their public access.

But luckily I got there early. Once inside, I overheard scientists discussing a scare at a local water facility, where a motorboat had breached security in the dead of night. I was there when water samples arrived to test for contamination. I watched as Dr. Wong ruled out a range of pathogenic germs on the list of suspected bioterror agents. Seeing her conduct the lab tests and grow the cultures of germs, each with its particular color and smell, was a revelation to me. The story demystified a frightening process and put readers at the center of local action. I didn't feel prescient at the time; I was just trying to work on a local medical angle to the 9/11 terror strikes. This just shows the value of an early call.

When an epidemic breaks, getting through the crush of media calls to authoritative sources can be frustrating or even impossible. But there are ways to speed up the sourcing process. Doing the preliminary legwork to read up on a disease can help you ask smart questions that invite a source to engage with you and give you the best possible interview.

A word about sourcing: It takes time to cultivate relationships. But the best reporters I know invest a lot of time doing this. It's especially valuable in science, where the subject matter is highly technical. One really compulsive reporter was famous for not simply collecting business cards, but noting on the back of each one where he met the expert and what they talked about on that occasion. That's super calculating. But it helps to refresh the memory of overscheduled experts if you remind them that you met at last year's World Bugs and Drugs Conference, at a session on flesh-eating bacteria, and now you are doing a follow-up story, and so on.

When covering new drugs, it's axiomatic that if a drug sounds too good to be true, it probably is.

Whenever you start to report a story about a promising new treatment, a useful piece of advice is not to ask "Is there any toxicity?" but rather to ask "What is the toxicity?" Any drug that is potent enough to be active in the body will, given a high enough dose, have some side effects somewhere in the body. So even when a source claims a drug is 100 percent safe, press for details: Is it toxic to the liver, kidney, bone marrow? And at what dose? This avoids a story that generates false or naive hope of a miracle cure.

While reporting on therapeutic research, reporters can be vulnerable to manipulation. Companies that are struggling to raise cash in the midst of a make-or-break clinical trial may offer you a peek at selected data, or a profile of their star patient who underwent a miraculous recovery. Beware of carve-outs of data that omit the inconvenient fact that nobody else in the study responded as well. Single-patient responses are anecdotes. Anecdotes don't prove a drug's worth; overall patient responses do. So if you use anecdotes, use them with care. Balance the promise with the potential for harm. Offer readers the caveat that it's impossible to draw conclusions from a single case—however compelling.

A word about profiles: The field of infectious diseases is brimming with human interest stories, from the perspectives of the patients, their families, and the researchers and doctors who investigate and treat their diseases. But a common pitfall for reporters is when they become so taken with the subject they are profiling that they end up with a puff or tribute piece. This happens by failing to frame the drama with enough objective background to make the story credible. Again, it's all about balance.

Resources abound: Most major health organizations have encyclopedic websites that are regularly updated with news and statistics on the whole universe of diseases and treatments. Here are some of the most useful:

The World Health Organization, based in Geneva, Switzerland, keeps encyclopedic global health statistics on epidemics at www.who.int.

The Joint United Nations Programme on HIV/AIDS (UNAIDS) posts its annual update of global statistics on the epidemic at www.unaids.org.

The U.S. Centers for Disease Control and Prevention in Atlanta runs a voluminous site with everything from flu statistics to travel medicine at www.cdc.gov. CDC also has a related bioterrorism site on germ warfare countermeasures and emergency preparedness at www.bt.cdc.gov.

Michael Osterholm of the University of Minnesota, a recent adviser to the federal government, runs a good infectious diseases website at www.cidrap.umn.edu.

The National Institute of Allergy and Infectious Diseases, a unit of the National Institutes of Health in Bethesda, Maryland, runs a helpful site at www.niaid.nih.gov.

When doing background on a field of research, I rely on Medline and other journal databases, accessible through PubMed, a service of the National Library of Medicine, which carries decades of research reports in abstract form. You can reach it by doing a Web search of the keyword "PubMed" or by going to www.ncbi.nlm.nih.gov/PubMed/.

Many professional organizations, such as the American Public Health Association (www.apha.org), have useful sites as well.

The Web also holds a universe of popular and special interest information that's useful for spotting trends in consumer health behavior, as well as for color and human interest. But not all sites provide authoritative data. Read the content with a healthy dose of *caveat lector.*

Consider adding annual infectious disease conferences to your calendar, such as the ICAAC (the Interscience Conference on Antimicrobial Agents and Chemotherapy). There's also the annual Conference on Retroviruses and Opportunistic Infections (CROI), and International AIDS Conference meeting every other year.

Among winter science meetings, Keystone Symposia stands out for cutting-edge science. Meetings cover topics ranging from AIDS to vaccines, and can offer education by immersion, plus an opportunity to meet researchers at poster sessions and to chat informally in the corridors. It's a great way to enrich your knowledge, your list of story ideas, and your Rolodex of sources.

After covering health for two decades at the *Journal*, I still can't imagine a more dynamic and engaging field than infectious diseases. I'll never forget what one man with HIV confided to me. He said that he imagined that the day the world cures AIDS will be like the celebration in the famous photograph of the end of World War II: Parades will wind through city streets, confetti will rain down, and strangers will embrace. That man didn't get to see his dream come to life, but I hope I'm there when the story breaks.

25

Nutrition

SALLY SQUIRES

Sally Squires is an award-winning medical and health writer for the *Washington Post,* where she also writes the nationally syndicated column the Lean Plate Club for the Washington Post Writers Group. She hosts the Lean Plate Club online Web chat every Tuesday at www.washingtonpost.com. Sally has covered science in Washington since 1981, beginning as the national health and medical correspondent for Newhouse Newspapers. She moved to the *Post* in 1984 to help start the Health section and serves as a regular television and radio commentator about health. She holds two master's degrees from Columbia University: one in journalism, the other in nutrition. The author of *Secrets of the Lean Plate Club* (2005) and co-author of *The Stoplight Diet for Children* (1987), she is the first journalist to be named an honorary fellow of the Society for Public Health Education, and the first to receive a journalism award from The American Society of Nutritional Sciences and The American Society of Clinical Nutrition.

Twenty years ago, if someone had suggested that nutrition news would regularly make the "A" section of major newspapers—and often the front page—I probably would have laughed. Sure, through the years, the occasional nutrition or weight-related story has made it to the front page. In 1998, a committee convened by the National Heart, Lung, and Blood Institute changed the definition of "overweight." Twenty-nine million Americans went to sleep thinking they were fine and woke up to learn that the government now said they needed to shed 6 to 12 pounds to be at a healthy weight. That story, which I covered for the *Washington Post*, made it not to just to the front page but above the fold.

The straight news story began this way:

> The federal government plans to change its definition of what is a healthy weight, a controversial move that would classify millions more Americans as being overweight.
>
> Under the new guidelines, an estimated 29 million Americans now considered normal weight will be redefined as overweight and advised to do everything they can to prevent further weight gain. Those who are already experiencing health effects, such as high blood pressure, elevated cholesterol or diabetes, will be encouraged to lose small amounts of weight—about six to 12 pounds—to bring them back to safer weight levels.

But in a follow-up piece that I did for the Health section, I had a little more fun with the off-the-news lead and wrote this:

> What do Olympic gold medal skier Picabo Street and Baltimore Orioles third baseman Cal Ripken Jr. have in common?
>
> According to new federal guidelines, they are both overweight.
>
> So what should they and the millions of other adults suddenly classified as overweight do about their extra pounds? Athletes like Street and Ripken may be special cases, but what about the rest of us?

In this follow-up story, I was able to offer a more in-depth explanation of the body mass index—a screening measure for determining a healthy weight that has replaced the old Metropolitan Life Insurance height and weight charts, which used body frame size, height, and gender to offer healthy body weight. That's because the weekly Health section often has more space—and, depending on the news cycle, more time—to digest the latest findings and provide more explanation to readers.

Nutrition news really began to heat up in 2000. One of the first stories to open the pipeline to the front page was based on a report from the National Academy of Sciences. It found that megadoses of the widely used antioxidants vitamin C, vitamin E, beta carotene, and selenium were ineffective at reducing the chances of getting cancer, heart disease, diabetes, Alzheimer's disease, or other illnesses. The widespread popularity of antioxidant dietary supplements and the fact that the *Post* had the story first helped to move this piece out front and nearly got me a live appearance on the *NBC Nightly News* with Tom Brokaw. (The *Washington Post* has an operating agreement with NBC News, so reporters often find themselves getting interviewed on hot topics.)

It all began because in the course of reporting, I had learned that the Food and Nutrition Board of the NAS's Institute of Medicine was about to issue a new report concluding that the common practice of taking large doses of vitamin C and other antioxidants was unnecessary. NAS reports are usually tightly held and have strict embargoes. But since I got this tip before ever receiving any information from the NAS—and confirmed it with multiple other sources—we went with the story. To make sure that I didn't break any embargoes—or even have the appearance of doing so—I also studiously avoided getting any materials from the NAS. I had to call them, of course, to get a comment. That's when the bargaining started. They asked me to delay. I said no. We called back and forth until finally the NAS simply broke their own embargo and hastily arranged a telephone press conference to issue the report.

That knocked me off the nightly news, but we still went with the story for the front page, saying:

> There is no convincing scientific evidence that taking large amounts of vitamin C, vitamin E, or the nutrients selenium and beta carotene can reduce the chances of getting cancer, heart disease, diabetes, Alzheimer's disease, or other illnesses, a National Academy of Sciences panel announced yesterday.
>
> Despite popular belief that high doses of these so-called antioxidants can protect the body from a variety of illnesses, including the common cold, there is insufficient evidence to recommend that Americans get more of these nutrients than is necessary to prevent basic nutritional deficiencies, the panel said. In fact, extremely high doses may lead to health problems rather than confer benefits, according to the panel, which for the first time set upper limits for vitamins C and E and for the mineral selenium.

Since then, it's not unusual to find nutrition and obesity-related stories earning front-page placement, even sometimes snagging the most coveted above-the-fold positions. The Diabetes Prevention Program findings that simple lifestyle changes—eating less, shedding a few pounds, exercising more—could prevent diabetes in more than half of people on the cusp of getting the disease were reported on page one. So were the new guidelines on controlling blood cholesterol levels with a combination of improved diet, more exercise, and cholesterol-lowering drugs.

The U.S. Department of Agriculture's review of popular diets was a slam dunk for our front page. Nothing but readers there! Same thing for a story on new body-weight charts for children to help combat childhood obesity. A Harvard School of Public Health report on the health benefits of the Mediterranean style of eating also earned a spot on page one. So did a story on how federal policy makers were pursuing legislative solutions modeled after anti-smoking programs to curb the obesity epidemic.

Small wonder. What people eat—as well as how much—is hot news in a nation where two out of every three adults are overweight. Forty-four million people are at least 30 pounds or more above their healthy weight, placing them squarely in the obese category. Another 123 million are overweight and headed for obesity if they don't do something to stop it.

Weight-related diseases add up to more than $200 billion annually in medical treatment and lost productivity. Some 112,000 people in this country die every year simply from being too fat. And obesity rivals smoking in cutting years off lives, because it fuels heart disease, cancer, stroke, and diabetes—four of the top 10 causes of death in the United States.

Nor is this growing girth simply a U.S. problem. In March 2003, the World Health Organization noted that weight-related chronic diseases have overtaken infectious diseases worldwide in mortality and morbidity. In a presentation at the 2004 Experimental Biology meeting in Washington, Eileen Kennedy, dean of the Tufts University School of Nutrition (and a former deputy undersecretary of the U.S. Department of Agriculture) noted that some developing countries are waging simultaneous battles against undernutrition and overnutrition. One in 10 families in some developing nations, Kennedy said, has one member who is undernourished and another who is overnourished.

Of course, interest in nutrition extends beyond food consumption. Dietary supplements are a multibillion-dollar industry for a public that seems to prefer popping a pill rather than getting the essential vitamins and minerals they need from food. Thanks to the 1994 Dietary Supplements Health and Education Act, there's virtually no oversight on daily vitamins and minerals or botanicals, such as St. John's wort.

There's debate about genetically modified food. There's concern about food safety, from mercury and PCBs in seafood to antibiotics in chicken and hormones in beef. Should we eat locally or globally? Organic or pesticide free? Whole grains or fortified? In short, nutrition news is a flavorful, often complicated topic.

While only a handful of major news organizations—the *Washington Post*, the *New York Times*, the *Wall Street Journal*, *USA Today*—have the resources to dedicate staff to covering nutrition, plenty of media outlets put reporters on the nutrition beat at least part of the time. Women's magazines and other specialty publications, including *Prevention*, *Men's Health*, and *Natural Health*, also spend a lot of column inches on nutrition.

So how does one report and write about a subject that touches readers' lives so intimately and yet seems to be constantly changing? One week, scientists report the benefits of drinking coffee. The next, its dangers. One year, fat is a four-letter word; in the next, the advice is to consume more healthy fat, such as that from canola, fish, and nuts. Not surprisingly, Food for Thought, a national public opinion poll conducted every two years by the nonprofit, food-industry-sponsored International Food Information Council, finds that the public is often confused these days about nutrition.

Here's what I've discovered in reporting about nutrition:

Make It Solid

Food and nutrition research has sometimes been considered a "softer" science. All the more reason to make sure that the stuff we dish out today is on rigorous footing. I start with a nugget of an idea: Does eating protein, for example, help boost metabolism? Then I log onto the National Library of Medicine's PubMed for a literature search. I download the abstracts, note the authors, and start calling or e-mailing them for interviews. If I'm checking out vitamins, minerals, or other dietary supplements, then the National Institutes of Health's Office of Dietary Supplements (ODS) is another good stop. There, I search through IBIDS, the ODS's online database of both peer-reviewed and non-peer-reviewed literature. For physical activity, I'll also start with a literature search on PubMed and then contact the American College of Sports Medicine for experts or cull through my own electronic database kept on Sourcetracker, a program designed by a former investigative reporter for news organizations. It conveniently links with Lotus Notes.

Keep Up to Speed

An increasing number of nutrition studies make the cut for the best-known journals: *The New England Journal of Medicine*, *JAMA*, even *Science*. But if you really want to know what's going on, don't miss the *American Journal of Clinical Nutrition*, the *Journal of Nutrition*, or *Journal of the American Dietetic Association*.

Across the Atlantic, the *European Journal of Clinical Nutrition* often features good nutrition morsels. Other good journals to keep tabs on include *Appetite* and *Obesity Research*. Some of the major scientific meetings to attend include Experimental Biology, sponsored by the Federation of American Societies for Experimental Biology (FASEB), and the annual meeting of the North American Association for the Study of Obesity.

A study published in the September 2003 *American Journal of Clinical Nutrition* provided a nice news peg for a Lean Plate Club column. Here's how it began:

> What exactly is it that vegetarians eat?
>
> That's a question asked by researchers in *The American Journal of Clinical Nutrition*, which has devoted much of this month's issue to the study of those who eat no meat, poultry, or fish.
>
> Or, at least profess that they do.
>
> It turns out that a number of people who report being vegetarians actually consume meat, poultry, and fish regularly. They just eat these foods less often than the rest of the nation's omnivores.

Ferret Out the Research Sponsors

Congress may have nearly doubled the NIH budget in recent years, but nutrition research is still catching up in funding after decades of neglect. So a lot of it winds up being sponsored by the food and dietary supplement industries. That doesn't necessarily affect the results, but at the *Washington Post*—as at most major news organizations—we generally make it a point to indicate when industry has sponsored research. And don't think that these connections between industry and scientists are limited to obscure, backwater universities. Some very well known researchers receive support from major industry groups. When in doubt, ask. Ditto for physical activity. When a new study recently crossed my desk about Curves, a nationally franchised gym aimed at women, one of our first questions was who sponsored the research. You can probably guess the answer.

Cast a Wide Net

Current nutritional advice is to eat a varied diet—in moderation, of course. That's a smart idea for reporting and writing about nutrition too. Whiffs of interesting new theories often come from unexpected places. So keep probing.

One example: the "fat virus." In a short item published in 2000, I wrote about an unusual line of investigation being pursued by Richard Atkinson, M.D., then at the University of Wisconsin—Madison: "As bizarre as it sounds, growing scientific evidence points to a common virus as a potential cause of at least a small percentage of human obesity. While the research has mostly been in animals, preliminary results from human studies appear to support the animal findings."

An updated story, published in July 2004, alerted readers to the fact that Atkinson had developed a blood test for the adenovirus 36, the so-called fat virus. Both stories also quoted scientists who could put the new theory in perspective and underscored that it still was just that: a theory, with some promising but still preliminary evidence to support it. Even so, we published the stories because they provided an unusual and new possible contributing cause to the obesity epidemic.

Keep It Fresh

Nutrition news isn't rocket science. Yet, both food and the other side of the weight equation, physical activity, are two areas where readers have plenty of personal experience to draw upon. Clever, catchy writing makes any subject more palatable, but it works especially well for stories about nutrition and physical activity that can otherwise seem stale.

Add a Dash of Skepticism

In 2003, a visiting scientist at the Harvard School of Public Health captured headlines when she reported what sounded like a dieter's dream at the annual meeting of the North American Association for the Study of Obesity: Low-carbohydrate eaters could consume more calories—300 more per day than low-fat dieters—and lose the same amount of weight. That provocative finding suggested that low-carb diets were somehow speeding metabolism and got a lot of media play. But the findings, which had not gone through the peer review required for publication in a scientific journal, were not statistically significant, and as of this writing, the study still had not been accepted for publication by a scientific journal.

Find as Much Consensus as Possible

My former science-writing professor at the Columbia University Graduate School of Journalism, Ken Goldstein, taught me this. Writing for *Reader's Digest* helped reinforce it when my editor and fact checker insisted that everyone quoted in the pieces had to pretty much agree with what everyone else said.

Sure, it's fine to report on new avenues of research—does increasing calcium intake really help boost weight loss?—but you must be sure that there are more than one or two scientists who are touting the theories. Expert groups convened by the federal government or major national organizations (the National Academy of Sciences' Institute of Medicine, the American Heart Association, the American Diabetes Association, the American Dietetic Association, the U.S. Preventive Services Task Force, the Cochrane Collaboration) generally reflect broad, evidence-based scientific consensus.

The reports they produce may sometimes be bland, but their ingredients have usually been well chewed by a large group of experts. So whether it's the National Academy of Sciences Dietary Reference Intake on how many vitamins, minerals, proteins, and carbohydrates we need to eat daily, the National Heart, Lung, and Blood Institute's advice on sodium or obesity, or the U.S. Dietary Guidelines and the Food Guide Pyramid, you'll find solid, scientific consensus—just what the public hungers for.

26

Mental Health

PAUL RAEBURN

Paul Raeburn is the author of *Acquainted With the Night: A Parent's Quest to Understand Depression and Bipolar Disorder in His Children,* which was published by Broadway Books in 2004. He was a senior editor and writer at *Business Week* until May 2004, and before that he was the science editor and the chief science correspondent at the Associated Press. Paul has written for the *New York Times, Popular Science, Child, American Health, Technology Review,* and many other newspapers and magazines. He is also the author of *Mars: Uncovering the Secrets of the Red Planet* (1998) and *The Last Harvest: The Genetic Gamble That Threatens To Destroy American Agriculture* (1995). Paul, a past president of the National Association of Science Writers and a recipient of its Science-in-Society Award, is a native of Detroit and a graduate of the Massachusetts Institute of Technology. He lives and works in New York City with his wife, the writer Elizabeth DeVita-Raeburn.

It wasn't until I had a profound personal experience with mental illness in my family that I started covering psychiatry and psychology. In the late 1990s, my son, Alex, was diagnosed with bipolar disorder. A few years later, my daughter, Alicia, began suffering from repeated bouts of severe depression. Even after they became ill, I resisted turning my reporting to mental health. But as I continued to experience the suffering that these illnesses can cause, I finally succumbed. If I was going to help my children, I needed to learn a lot more about psychiatry, both research and treatment.

With a background covering research, I could have confined my reporting to published studies and conferences, the bread-and-butter of science coverage. But I quickly realized that by taking that approach, I would be getting only a small piece of the story. For one thing, research in the behavioral sciences is, as I had always suspected, at a rather primitive stage. Researchers know far more about the heart, the kidneys, and tumor cells than they do about the brain. That's understandable; the brain is a far more complex organ.

The scandal, however, is that what is known about the brain is rarely taught to psychiatrists. "Most of the more advanced training for psychiatric residents is really apprenticeship training in which brain science plays little or no part," write the Harvard psychiatrist J. Allan Hobson and the writer Jonathan A. Leonard in their book, *Out of Its Mind: Psychiatry in Crisis* (2002). "The brain science knowledge of many practicing psychiatrists remains mostly informal or even anecdotal, leaving psychotherapy and psychopharmacology separated, isolated, and diminished at a time when brain science has the ability to nourish and combine them in an empowering fashion."

The message to reporters is that if we are to understand psychiatry, psychology, and mental illness, and write capably about them, we must do more than peruse the scientific journals and attend the neuroscience meetings. We need to get out there in the trenches, by which I mean the homes and the minds and hearts of the families who are suffering from mental illness.

You know these people already, although you might not be aware of it. Since I began writing about my experiences with my children's mental illness, I have discovered dozens of friends, acquaintances, and colleagues who are grappling with the same thing. People I'd known for 15 or 20 years would say, "What are you working on?" And when I told them, they would often respond with a story about one of their own children, or a parent, or a brother or a sister who suffered from mental illness. Rarely do these stories have happy endings. In many cases, the families are still suffering, still searching for something to help the schizophrenic brother who has been in and out of institutions for 20 years, or the child who has been failing in school and treated with psychiatric drugs for several years but still lacks a diagnosis—and isn't getting better. Sometimes it's a story about the loved one who suffered so intensely that he or she could no longer face the future, and chose not to go on living.

Certainly, it is important for us, as science writers, to pay attention to the science. Psychiatric research rarely makes it into the pages of *Science* or *Nature*, or the *New England Journal of Medicine*. *JAMA*, alone among the journals most closely watched by science writers, does publish psychiatric studies regularly, and they are often among the best. The American Psychiatric Association is the principal source of news on mental health. Many of the studies reported at its annual meeting are small and of dubious quality, and much of the meeting is devoted to continuing education for psychiatrists, not the latest research. But by paying careful attention to the meeting's program and abstracts, you will find out about most of the important clinical trials.

What you will miss, of course, is the perspective of real patients in the real world. In 2003, questions were raised about whether certain antidepressant drugs, including Prozac, Paxil, Zoloft, and others, could increase the risk of suicide in teenagers. Talking to officials at the Food and Drug Administration and to the authors of studies on these drugs is not enough. In my reporting, I talked to parents. Many news stories seem to accept that parents are much too willing to dose their kids with drugs, grasping at any chance to give them an advantage at school or at home. I found the opposite. Parents agonized over the decision to put their youngsters on brain-altering medications. The new questions being raised about a possible suicide risk deepened their agony. The result was a personal story that conveyed a message quite different from any of the other stories I read. I wrote it as a commentary for NPR's *Morning Edition.* It began by recalling my daughter's suicide attempt, and my decision to put her on an antidepressant. And it continued:

> Now, evidence is emerging that perhaps antidepressants themselves can trigger a suicide attempt. Millions of children are taking them, and there isn't a single study that can tell parents whether the risks outweigh the benefits. Doctors and government regulators make educated guesses, but they don't know anything for certain. Research studying the risks of antidepressants on children has never been done.
>
> For now, my daughter has continued on antidepressants, and she is doing much better after a few difficult years . . . I did what most doctors recommend—I kept her on the drugs, fearing that without them, her depression could roar back, putting her at far greater risk . . .

For news of basic research on the brain, the best place to go is the website and the annual meeting of the Society for Neuroscience. If you compare the news coming out of the Society for Neuroscience with the news coming out of the American Psychiatric Association, you will soon see the split that Hobson and Leonard decry in their book. These two groups do not speak the same language. The neuroscientists rarely consider the implications of their work for mental illnesses, and the psychiatrists, except those involved in research, don't understand the jargon and premises of neuroscience well enough to grasp its import. That means, of course, that this is a wonderful opportunity for reporters to bridge that gap with stories that bring together insights from psychiatry and from neuroscience.

The mental health story reaches far beyond the research news, however. In most health insurance plans, mental illnesses are covered at a lower level than other illnesses. The typical policy will reimburse psychiatrists and therapists at levels far lower than for other doctors. The number of allowable hospital days is usually less, and the co-payments for psychiatric drugs may be higher than for other drugs. There may be tighter limits on which providers can be seen. It's always important in medical research stories to address the issue of costs: Who can afford to benefit from newly developed drugs and diagnostic equipment, and how will they affect the nation's health care costs? But it's doubly important to raise these questions when reporting on psychiatric research, because of the tighter restrictions on coverage.

When I was at *Business Week,* I decided to write a story about proposals in Congress to require that mental illness and other illnesses be covered equally—the question of mental health parity, as it's called. Many stories had been written, but none answered what I thought was the central question: What would it cost to improve mental health coverage? Industry predicted dire effects on the U.S. economy, and health advocates said that the country had no choice but to bear those costs. But what were the costs? After much searching and reporting, I found a RAND Institute report—a cold, economic document—that said mental health parity would raise health care costs by about 1 percent. Here's how I used it:

> Opponents are again warning about costs, but new data and several independent studies suggest that employers' costs for mental health parity could be very small—amounting to less than a 1 percent increase in health care costs . . .
>
> One of the new studies was done by the RAND Corp. Based on an examination of 24 managed-care mental health plans covering 140,000 people, the study concluded that the added cost of providing mental health coverage equivalent to other medical coverage would come to less than 1 percent. The RAND researchers also looked at mental health care costs in Ohio, which instituted equal coverage a decade ago. "Their costs are totally stable—there is no big increase, no disaster," says Roland Sturm, an economist and the study's author.

And I didn't stop there. It turned out that mental health parity had recently been extended to federal workers. The increase in costs? About 1 percent. Clearly, mental health parity was not going to wreck the U.S. economy. But too many stories had allowed industry sources to say so—without any evidence to back up the claim.

If you're covering mental health, it's critically important to talk to patients, and their families, to get the complete story. And that can be a little bit tricky. The prospect might be a little frightening. It was for me, at first, because I had no idea what to expect. What I discovered was that many people with these ailments have a sophisticated understanding of them and are very articulate when discussing their experiences—as is the case with people with cancer, diabetes, or any other ailment. Of course, some do not have such a clear understanding of what's happening. They are no more or less intelligent or sophisticated than the rest of us. Indeed, they are us, in every way except one—they are sick.

Interviewing patients with mental illness and their families requires a degree of sensitivity that isn't required when talking to researchers. And it can raise difficult ethical issues.

During one recent interview, I asked the parents of a child who had been diagnosed with bipolar disorder to describe their child's symptoms. They were experts on the subject of their child's illness, of course, and they went on at great length. Their child, like many children, had seen numerous psychiatrists and therapists before being given the diagnosis of bipolar disorder. They had now settled into a new treatment regimen, which was only partly working. Their child was still having difficulty in school and at home. As part of the story, I interviewed several researchers who study bipolar disorder in children. When I described the child's symptoms given by the parents, one of the researchers hesitated for a moment on the phone and then said, "I don't think this child has bipolar disorder."

If that was the case, the child was getting the wrong treatment. This presented an ethical dilemma I hadn't faced before. Should I tell the parents? I worried about this for a week or two, and then I called the researcher again. I explained the problem, and I asked him what I should do. If I was going to tell them that their child was misdiagnosed, I wanted to be able to refer them to this researcher—who lived a thousand miles away—to find out what to do. The researcher said that it was best to leave things alone. There was little that he could do to help, and the child was at least partly responding to the treatment. "They have a program in place, and we shouldn't disrupt that," he said. I wasn't happy with that outcome. If I were the parent, I would want to know. But it wasn't that simple. The researcher, for one thing, was relying on my summary of what the parents had told me. He had never examined the child. He couldn't make a definitive diagnosis on that basis. So I didn't say anything to the parents. And I don't know whether that was the right thing to do. I worry about it still.

Some of the best sources of information on mental health are the voluntary advocacy organizations. These include the National Alliance for the Mentally Ill (www.nami.org), the National Mental Health Association (www.nmha.org), and specialty groups, such as the Child and Adolescent Bipolar Foundation (www.bpkids.org), the Depression and Bipolar Support Alliance (www.dbsalliance.org), and many others. Many of these groups hold annual meetings, and they all have websites with valuable information. Many of them also include chat groups, some of which are open to the public. Paging through those postings is an excellent way to discover the concerns of patients and their families. Many of these groups can also provide experts and advocates to provide counterpoint to the researchers that you interview. NAMI and NMHA actively lobby Congress and the health care industry on behalf of people with mental illness, so they are a good place to turn for information on legislative proposals and insurance industry developments.

The National Institute of Mental Health is the government's primary sponsor of research on mental illnesses and the basic science underlying these disorders. Its website (www.nimh.nih.gov) includes press releases on recent research findings, along with short descriptions of the various kinds of mental illness. You can sign up for an email alert that will bring you news of new postings on the website, including new research findings. The American Psychiatric Association (www.psych.org) will refer you to experts both on research and on the problems with the health care system. The organization lobbies on behalf of psychiatrists, of course, whose interests often ally with those of patients, but sometimes don't. The American Psychological Association (www.apa.org) covers a far broader range of topics than mental illness, but it can also be a valuable source of information, and it, too, provides a referral list of expert sources.

You could fill several bookshelves, as I have, with reference sources useful for covering mental health. The single most important reference is the latest edition of the *Diagnostic and Statistical Manual of Mental Disorders,* otherwise known as *DSM*. The most recent edition, with the latest updates, is *DSM-IV-TR* (for "fourth edition, text revision") but as of this writing an earlier edition of *DSM-IV* will serve you just fine in most cases. It is the standard manual for the diagnosis of mental disorders. It's a curious thing, a sort of Chinese-menu list of symptoms, which, in drastic oversimplification, works something like this: Identify two symptoms from column A, three from column B, and one from column C, and it's anxiety disorder! The book exists because there is no diagnostic test for mental illness, so psychiatrists must rely on a catalogue of symptoms to make their diagnoses. It provides the official, widely accepted technical descriptions of the many kinds of mental illnesses, and while it might not quite allow you to make diagnoses yourself, it will certainly help you raise all the right questions when you're doing an interview.

There's one final thing to remember when writing about mental health. Many of the stories you will encounter are unspeakably sad and moving. But the outcome is not always bleak. My children are doing much better than they were a few years ago. They are getting good treatment, and they understand how to cope with their illnesses. I predict a bright future for them. And that will be true of many of the patients and families you meet in the course of your reporting on mental health. Sometimes sad stories turn out to have happy endings. If you want your story to be fair and accurate, save a paragraph or two to report that there's hope. Things do get better.

27

The Biology of Behavior

KEVIN BEGOS

Kevin Begos is an investigative reporter with the *Winston-Salem Journal* and was recently a Knight Science Journalism Fellow at MIT. He was the first reporter to gain access to thousands of sealed records from the North Carolina Eugenics Board, a state agency that ordered sterilizations of 7,600 people between 1929 and 1974. The *Journal's* series on eugenics won awards from Investigative Reporters and Editors, the Newspaper Guild, and the Society of Professional Journalists. Kevin was previously the Washington, D.C., correspondent for the *Journal,* covering politics and policy issues (especially the environment), and he has also done reporting from Iraq, Afghanistan, Pakistan, Sudan, the West Bank, and other countries. He is working on his first book, about eugenics in post–World War II America.

In 1970, Nobel Prize winner William Shockley made a dramatic declaration: that the average IQs of black people were significantly lower than those of whites, and that blacks of low intelligence should be paid by society to be sterilized.

Shockley's Nobel was for work he conducted at Bell Telephone Labs that contributed to the discovery of the transistor. He was not an expert in genetics, biology, sociology, or anything to do with the human mind, behavior, or reproduction. Yet he was able to use his status as a "Nobel laureate" to get vast amounts of media coverage for his sterilization plan.

Why did journalists give Shockley so much ink? Would they—or their editors—send a troubled child who needed help to a TV repair shop, or send a broken computer to the office of a psychologist at Harvard University? Why, then, would they quote a physicist like Shockley when writing about race and intelligence?

The subject of the biology and genetics of behavior raises many questions like these. It is a fascinating field to write about, but it will take you into some pretty tricky terrain. You'll often find yourself (and your sources) moving back and forth across two vastly different scientific domains—the laboratory, which has traditionally been based on chemistry, biology, and experiments that can be duplicated and proven, and theoretical science, which aims to uncover and explain broad concepts about life. The people you encounter will have specific areas of expertise, but some may (consciously or not) attempt to make grand statements about how a particular idea or discovery may affect humanity.

This is a huge, complicated, controversial subject just waiting to suck journalists into its hungry maw, from which it will spit us out in little pieces.

Okay, I'm exaggerating (a little). But it can be overwhelming to figure out even how to begin. There's Darwin and cell biology, psychology, sociology, religion, and politics. There are historical figures such as B. F. Skinner and current stars such as Noam Chomsky at MIT and Harvard's E. O. Wilson and Steven Pinker. And there's the whole issue of racism at the edges. Do you have to go into Hitler and the dangerous idea of a Master Race every time you write about behavioral genetics?

The short answer is: It depends. What's your deadline? What's your word count? There's a huge difference between an 800-word spot newspaper story you have to do in a day and a 1,500-word feature you have a week to write, or a longer magazine piece that may take a month or more. It's easy to get carried away researching and writing about genetics and the biology of behavior. That's good in an intellectual sense but something to keep tabs on as a journalist. The type of story you're doing—long or short, newspaper or magazine, local or national—affects every other decision you'll make.

The complexity of the subject can be a blessing. There is so much research going on that it's likely you can find a creative way to address many issues. Say you're interested in brain scans, vision, and human emotions. Consider this lead from a March 2004 *National Geographic News* article by Stefan Lavgren:

> Do we all see the world in the same way? To answer that age-old question, a group of Israeli researchers went to the movies. Using magnetic resonance imaging (MRI), the scientists monitored the brain activity of volunteers as they watched the classic Clint Eastwood Western *The Good, the Bad, and the Ugly*.

I'm not going to tell you how that study turned out, but I bet you're curious—and your editor would be, too.

Or say you've found a researcher who's looking at the behavior of violent criminals and ways to rehabilitate them. You could pair that scientist's research with interviews of people in prison, and perhaps even with victims of crime and their families. Then, instead of a story that's just about whether science could, in the future, "cure" criminals of violent tendencies, you have the personal and intense issue of whether victims and society would even accept and make use of such a scientific breakthrough.

There are many critical "should we" issues in science. Should we learn more about the behavioral components of genetic differences, or brain structure and function? Advanced genetic tests will give parents (and perhaps insurance companies) the chance to consider many more aspects about the makeup of a child before birth—should such tests be allowed? Brain scans are giving increasingly detailed pictures of brain activity—should such evidence be introduced in court? Will the police of the future be able to make a person take a brain scan or a DNA screen the way fingerprints or mug shots are done now?

These are the kind of ethical issues that make writing about the biology of behavior fundamentally different from writing about, say, a new ultra-light, ultra-strong material, or a breakthrough computer chip design. In hard science it's going to be possible—even relatively easy—to find a knowledgeable expert to say X is or isn't a real potential breakthrough or solution to some issue that previously stumped everyone in the field. There will be some differing views on any new science story, but you should be able to find people to give you *reasons* for their views. For example, "It's not a breakthrough because we already knew something very similar from previous research," or "This is how it differs from previous research."

Finding an alternative point of view in a story on behavioral genetics is a little trickier. You need to be wary of any "expert" who claims a new study is a definitive breakthrough. Look for knowledgeable, level-headed people in the same field to give you feedback and quotes. Keep to some journalism basics, too. Find differing viewpoints on the issue, but only use those that seem to illustrate a broad consensus. If you're researching a controversial article that was published in a respected scientific journal and you find many people who are really (and specifically) critical, that's relevant. If knowledgeable people seem to be jumping on the bandwagon, that's relevant, too, as long as you remember historical lessons such as the stories about Shockley and race IQ. And if you really get stuck with a controversial story where expert opinion seems evenly divided down the middle, try to look hard at the science, and stress that this may be one of those issues that will need much more research before it's resolved.

Some scientists might not agree with me—some journalists might not, either—but I'd say that every controversial biology of behavior story needs to have a point of view up near the top that examines whether society would ever want to make use of some particular knowledge, even if we developed the ability to do so. One good thing here is that such points, if made relatively near the start of a story, can be effective even if they're brief.

Let's put these rules into play in a hypothetical scenario. Imagine that you learn that incredibly important and controversial research will be published in two days. The research strongly suggests that there are some fundamental genetic differences between blacks and whites that lead to differences in, say, sexual behavior. It's going to be big news—you have to write about it, but you don't have much time.

The fact is, you can't count on getting Steven Pinker or E. O. Wilson on the phone on short notice—but you also may not want to. If you're writing for a regional or niche audience (a southern newspaper, or a publication for social workers), consider adding someone to your story who will look at the science from the perspective of real-world society. That doesn't mean you get a social worker to comment on the science in the study, but rather on how the science might be used—or misused. That local (or specific) angle gives you two things: someone your readers should be able to relate to, and a view that goes beyond simply stringing together different scientists in a "he said, she said" battle over whether the research is correct.

One little thing to watch for: When scientists tell you about some amazing research they're doing that they "plan to publish" in the future, in almost all cases you should wait, at the very least, until the results have actually been accepted by a reputable journal. Peer review isn't perfect, but it does serve a role. If you quote a lawyer who "plans to file a lawsuit," you're giving publicity to a threat, not law. The same rule applies to science. Talk is cheap.

After you've written your genetic, race, and sexuality story on a frenetic two-day deadline, you're probably coming to see that the subject—like many aspects of biology and behavior—really deserves more space and time. But how do you convince an editor to let you really examine the issue in a long piece that isn't about a breakthrough treatment for cancer or some other disease?

It helps to find a unique angle. Look at Anahad O'Connor's treatment of a related subject, a story on the behavioral effects of serotonin, in a May 2004 article in the *New York Times*. O'Connor quotes Helen E. Fisher, a Rutgers University anthropologist and an expert on the biology of romantic love:

"We know that there are some real sexual problems associated with serotonin-enhancing medications" such as antidepressants, Fisher was quoted as saying. "But when you cripple a person's sexual desire and arousal, you're also jeopardizing their ability to fall in love and to stay in love."

The article noted that Fisher and a colleague studied the brains of people in love and that "Lust is fueled by androgens and estrogens. Attachment is controlled by oxytocin and vasopressin. And attraction, they say, is driven by high levels of dopamine and norepinephrine, as well as low levels of serotonin."

O'Connor ended the article with a kind of qualifying sentence that I think should be present in some form in almost every article about the biology of behavior: " 'Everyone is distinctly different,' Dr. Fisher said. 'Some people are so securely attached that this isn't going to change things for them. But people should be aware that these drugs dull the emotions, including the positive ones that are central components of romantic love.' "

That reminder—that the biology of behavior is ultimately about *individuals*—brings us to some important ethical issues to remember. There's a sad history of scientists, at times, misusing their positions of authority to promote racist, sexist, or other discriminatory behavior—and of journalists unquestioningly going along for the ride.

Consider this story from the *Washington Post* from early in the last century:

> SAVANTS AND EUGENICS
>
> Noted Men Cooperate in Plan to Improve Human Race.
>
> NATION-WIDE LAWS SOUGHT
>
> Dr. Davenport, the New York Biologist, Declares Nation-Wide Segregation and Curbing of Defectives Will Make Perfect Manhood and Womanhood—Mrs. E. H. Harriman Gives Financial Aid.

That 1915 story was about a plan to sterilize massive numbers of "defective" people, and the head of the committee was Alexander Graham Bell. Over the next few decades there was an enormous tide of such sentiment in America. The press was right there, often assuring readers that the best "science" supported eugenics—the idea that humanity could be bred as if it were livestock or crops, culling the weak and breeding the strong. Professors from Harvard, Yale, Princeton, Cal Tech, and other leading schools were at the forefront of the eugenics movement. They received funding for their research from the Rockefeller and Carnegie Foundations, and their papers were published by the American Association for the Advancement of Science and leading journals.

Hitler and other Nazi leaders made no secret of the fact that programs that allowed the forced sterilization of hundreds of thousands of Germans in the 1930s were directly based on American laws, as historian Stefan Kuhl has noted. When California eugenicist C. M. Goethe visited Germany in 1934, he wrote back to his colleague Eugene S. Gosney, "You will be interested to know that your work has played a powerful part in shaping the opinions of the group of intellectuals who are behind Hitler in this epoch-making [sterilization] program. Everywhere I sensed that their opinions have been tremendously stimulated by American thought . . ."

Most scientists (and journalists) backed away from supporting eugenic sterilization after World War II. But not all.

"The danger is in the moron group which includes a host of physically attractive individuals whose IQs are lower than a January thermometer reading," wrote journalist Chester Davis in a full-page Sunday feature for the *Winston-Salem Journal and Sentinel* in 1948. "Among other things, they breed like mink." The headline blared "The Case for Sterilization—Quality Versus Quantity."

The publication of *The Bell Curve* in 1994 showed that academic racism is still alive, and though such views are in the minority among scientists today, other developments suggest journalists need to be even more careful when covering the revolution in genetics.

"The ethical issues that were raised by eugenics are likely to be the very same ethical issues that are being raised with genetic research, now and in the future," said Steve Selden, a historian at the University of Maryland. "They didn't have the technology to achieve their goals," he said of the eugenics movement of the early twentieth century. "We do."

Harvard biologist E. O. Wilson agrees, writing in his book *Consilience* (1998) that within the next 50 years it's likely that humanity "will be positioned godlike to take control of its own fate." That prospect, he says, "will present the most profound intellectual and ethical choices humanity has ever faced."

It's up to us as journalists to examine and explain not just the science and technology behind this amazing genetic revolution, but the people and the ethics, too.

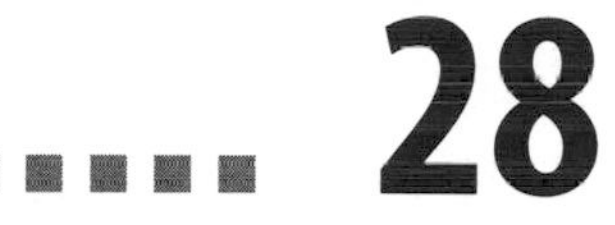

28

Human Genetics

ANTONIO REGALADO

Antonio Regalado is a science reporter at the *Wall Street Journal*. Before joining the Journal he was an editor at MIT's *Technology Review* magazine from 1998 to 2000, where he was involved in coverage of biotechnology, robotics, and patent issues. From 1995 to 1998, he was a staff writer at Windhover Information, in Norwalk, Connecticut, a publisher of trade magazines covering the drug, diagnostics, and medical device industries. His reporting on the rationing of medicine was part of a series that won the National Institute for Health Care Management Foundation journalism award in 2004 and was a Pulitzer finalist.

Genetic research is moving faster than a nematode poked by a platinum needle. Every week, the scientific journals report a score of new gene discoveries made in mice, worms, and men. How can a science journalist cover it all?

It's hopeless, of course. So one thing I always keep in mind is it's often the *methods* or *scientific tools* behind these molecular discoveries, not the discoveries themselves, that present the best story possibilities. Examples of topics for such "tool stories" include DNA chips, proteomics, and new imaging technologies like the green-fluorescent protein used to make zebrafish and other laboratory critters glow.

In writing about the technologies that drive biological research, I've found a formula that has worked well for me, time and again. Of course, not every story fits the same mold, and the best ones break it. But it's important to be familiar with how a tool story typically comes to be, and how to write one.

I like to think about biology as a big onion that's rapidly being peeled. There are tens of thousands of biologists peeling away every day, figuring out all of life's working parts. But I never saw much sense in inspecting every peel for its news potential. (And some editors I know refer dismissively to the latest uncovering of a gene for heart attack or schizophrenia as "gene-of-the-week" stories.) It's better, sometimes, to focus on the new techniques and ideas for peeling the onion.

Tool stories are big-picture stories that can be newsy, but the trends tend to have a long shelf life. They endure through numerous news cycles, and ultimately nearly every outlet in the journalistic food chain will cover the big ones. Your decision is when to catch the wave. Some reporters put a big emphasis on being first, but others will be content to watch the story unfold and cover their piece of it when it's right for whatever market they happen to be writing for.

Either way, a tale of how a new technology is changing biological research is a great way to teach your readers—and yourself—about how science really works. From high-powered gene-sequencing machines to cloning to the latest advances in biomedical imaging, it's very often new technology that determines what research gets done, as well as the agenda of funding agencies like the National Institutes of Health (NIH).

For instance, did you know that way back in the 1980s some scientists proposed an ambitious effort called the Human Protein Project to map all human proteins? It never happened. Instead, the NIH backed the Human Genome Project for one big reason: Proteins were tough to study, while genes were far easier to sequence. The tools dictate the science.

Gene-sequencing technology is probably the best example of the tool phenomenon. The technology has evolved to a point where laboratories full of humming machines spew out data on a 24-hour basis. Gene-sequencing technology spawned the genomics industry. It changed science and changed our ideas about how to write about genetics. The story isn't that scientists have found a gene—now they find them by the bushelful. The story is how they found it.

Readers aren't very interested in machines, or in the details of how a technology works. That sets a limit on how much detail can go into a story and may send you hunting for a better vehicle for your tale. Luckily for you, there are always people behind these inventions, sometimes very interesting ones. Think of Kary Mullis, the quirky surfer dude who invented the polymerase chain reaction (PCR). Or better still, J. Craig Venter, the business-minded biologist who founded Celera Genomics, the company that sequenced the human genome.

Venter's sharp tongue and impolitic ways helped turn the genome project into a public-private race where the kind of ambitions and jealousies that always infuse science spilled out into public view. Although the focus of journalists was mostly on the competition and on Venter's personality, behind Celera's bid was a technological advancement—a new machine with 96 thin glass capillaries that could sequence much faster. An article in *Time* in January 1999, by Michael D. Lemonick and Dick Thompson, began this way:

> When the Human Genome Project was launched a little under a decade ago, boosters compared it with the Manhattan Project or the mission to put men on the moon: an effort so complex and so broad in scope that only the government had the financial and bureaucratic resources to pull it off . . .

The article went on to describe the shock of public-sector researchers at the founding of Celera and similar efforts, which not only brought in the "profit motive" but had also had "found ways to speed up the decoding process." A new tool was about to strip the peel off the Human Genome onion in record time.

Looking back through the clips from this period, covering roughly 1998–2001, I find hardly any stories that actually mention the word "capillary." Even fewer discuss the competition between Perkin-Elmer Corp. (the company that bankrolled Celera and made off with hundreds of millions in profits when its stock shot up) and rival Amersham-Pharmacia Biotech to create such instruments. Almost none explain that the only reason the public project managed to keep up is that its members started buying the advanced sequencers, too.

I think some important aspects of a great story were largely missed. But journalists understandably focused on the bitter competition to complete the genome first and on Venter's high-IQ antics. In the pecking order of journalistic clichés, the tool story still ranks below "racing" or "crazy genius" stories.

■ ■ ■

Business magazines, alerted by venture capitalists whose investments are ready for publicity, are often first to note new technological trends in biology. For instance, writer David Stipp put DNA chips on the cover of *Fortune* magazine back in 1997, well before other mainstream publications were writing about them. The cover had a beautiful stained-glass image of a chip read-out, and the story's headline shouted: "Gene Chip Breakthrough—Microprocessors Have Reshaped Our Economy, Spawned Vast Fortunes, and Changed the Way We Live. Gene Chips Could Be Even Bigger."

Stipp's analogy between the computer revolution and DNA chips was powerful and adroitly framed. He was telling readers, quickly, that a new technology was about to transform biology in fundamental ways that they could understand:

> These world-changing biochips, formally known as DNA arrays, bear an uncanny resemblance to the chips that ushered in the information age. Instead of transistors, they are crammed with dense grids of molecular tweezers built to grip DNA. They give medical researchers the ability to analyze thousands of genes at once—in effect, to speed-read the book of life.
>
> Affymetrix, in Santa Clara, California, introduced the first biochip only last year. Yet the technology's startling implications are already coming into focus. Researchers at the company are using gene chips in landmark studies on everything from the origins of cancer to gene mutations that make the AIDS virus resistant to drugs. Exults Affymetrix cancer researcher David Mack: "I am so happy to be part of this."

I loved that last quote the moment I read it. At the time Stipp's article came out, I'd been writing lengthy articles about DNA chips for a trade magazine. Stipp's article helped open my eyes to the wider journalistic possibilities of the tool story, and in no small way accounts for how I ended up at the *Journal.*

One caution here: Some journalists got a little carried away with the computer-chip analogy. A number of stories (even in the *Wall Street Journal*) have reported that DNA chips are actually made out of silicon. In most cases, they're not. They're made of glass or plastic. If you're writing about a new tool, take the time to learn how it works.

Not all tools are as big a deal as DNA chips, but biologists are constantly inventing new techniques, and the more important ones tend to develop in predictable ways. I've observed this phenomenon pretty closely, and so here are a couple of clues for finding these stories.

Typically, the first step is the publication of the new method in a scholarly journal, often an obscure one. The implications may not be immediately clear, but if a technique works, then more scientists will want to try it, and companies will begin staking out a position. With commercial activity, trade publications such as *Chemical & Engineering News* or the news pages of *Science* are usually the first to note the excitement.

Reporters at general-interest outlets should be sure to read the trades. Even though they publish a lot of inside baseball, these publications remain a great place to find stories crying out for a wider audience. Even the help wanted ads in the back of *Science* or *Nature* can give you a clue as to what technology is hot.

Another way is simply to ask. Back in 2002 I was interviewing Phillip Sharp, the Nobel Prize winner who runs a brain science center at MIT. "What's the biggest story in biology right now?" I inquired.

Sharp's answer: RNA interference.

It turned out to be a classic tool story. RNA interference is a technology that allows scientists to precisely shut off any gene they wish inside cells, even inside living animals. The new method was being widely adopted in academia, and after a couple of calls to venture capitalists, it was clear to me that companies were also starting to get involved. I like to think the *Wall Street Journal* broke this one open for general audiences with a front-page story in August 2002, which began:

> Richard Jorgensen's idea was simple enough: Make bright purple petunias by splicing into the plants an extra copy of the gene that makes purple pigment. To his astonishment, the flowers bloomed white.
>
> That curious outcome defied genetic logic. After appearing on the cover of a prominent plant journal, the puzzling result prompted a wave of scientific inquiry. Now, more than a decade later, biologists are starting to get a handle on what went wrong in Dr. Jorgensen's lab and are calling the findings an important breakthrough.
>
> Scientists working on the petunia mystery have uncovered what is shaping up to be a critical piece of cellular machinery, a process by which plant and animal cells seem to blot out the activity of particular genes. Scientists say the discovery helps explain a lot that had perplexed them about life's basic functions, and they are already applying it as a research tool in the hunt for new medicines. Venture capitalists also are betting that it will yield super drugs that act like molecular torpedoes aimed at HIV or cancer. Scores of companies and academic labs have joined the hunt.

RNA interference went on to become *Science* magazine's Breakthrough of the Year, and many newspapers and magazines have printed big features about it. A story by *Fortune*'s David Stipp even called RNA interference "Biotech's Billion Dollar Breakthrough"!

Bees, Bugs, and Bio-Criminals

The story of a new tool in genetics research goes through cycles. At first, there's a big wave of coverage as everyone realizes the revolutionary potential of peeling that darn onion that much faster or better. Then the hype may fade, only to come roaring back when a well-established technology finds a new killer-application.

For instance, once the Human Genome Project was on its way to completion, scientists began turning their immense DNA-decoding capacity on other organisms. That turned the story of genomics in new directions, such as efforts to sequence the genome of every organism in an ecosystem. This inevitable shift in emphasis has in itself been a story for alert journalists. Laurie McGinley, who heads up health reporting at the *Journal*'s Washington bureau, did a page one story on February 28, 2003, that carried the *Journal*'s signature, multipronged headline emphasizing oddball details:

> Natural Selection:
> After Humans, Herds
> Of Animals Line Up
> For Gene Sequencing
>
> ---
>
> *Dogs, Chimps and Protozoa*
> *All Have Their Backers*
> *In a High-Stakes Contest*
>
> ---
>
> Designing the Perfect Bee

The story was about how the National Institutes of Health was about to "wrap up much of its work on the human, mouse, and rat genomes" and, with all its sequencing machines coming available, had launched a competition to see which organisms would be decoded next. The story featured lots of impassioned advocates for different species, including Danny Weaver, the "fourth-generation beekeeper" that Laurie used in her lead.

That was a fun story. But one of the most dramatic and significant changes brought about by genomics is our ability very rapidly to decode the genes of microbes, including viruses and bacteria. In a post–September 11th world, this sequencing ability has taken on unprecedented importance in dealing with biological threats, both man-made and natural.

For science writers, it all began with the deadly anthrax mailings sent to Capitol Hill and media organizations starting in October 2001. It looked like a perfect crime—the killer hadn't left any fingerprints on the envelopes and had snarled investigators from the FBI with a series of misleading clues, including fake return addresses.

But the anthrax mailer may not have counted on what gene-sequencing technology was capable of.

I happened to know that the microbe specialists at the Institute for Genomics Research (TIGR), in Gaithersburg, Maryland, had already been working to decode the anthrax genome for a couple of years. I suspected that decoding the strain sent by the mystery bioterrorist would be a priority, and that it could be done fast. A few phone calls later and the *Journal* had nailed the scoop: The National Science Foundation planned to give TIGR $200,000 to sequence anthrax bacteria isolated from the spinal fluid of Robert Stevens, the American Media Inc. photo editor who was the first of five people killed.

Everybody wondered, would the DNA of the killer strain match supplies of anthrax held by a particular scientist or lab? Suddenly it was science reporters at the *New York Times*, the *Washington Post*, and the *Baltimore Sun*—not their crime-desk colleagues—who began breaking news on the investigation.

The anthrax case broke open a new area—genomic epidemiology. Almost instantly, scientists can now decode a killer organism and uncover crucial clues about where it came from. And anthrax was only the beginning of the trend. With the first SARS outbreak, in 2003, and then with the spread of H5N1 avian influenza in Asia, gene sequences became available in record-setting time. The SARS sequence was complete in just a month, avian flu inside a week.

Who would have thought plain-old-vanilla gene sequencing would be on the front page again? When SARS was sequenced in April 2003, the *New York Times* put Donald G. McNeil Jr.'s story on page one:

> Scientists in Canada announced over the weekend that they have broken the genetic code of the virus suspected of causing severe acute respiratory syndrome.
>
> Sequencing the genome—which computers at the British Columbia Cancer Agency in Vancouver completed at 4 a.m. Saturday after a team slaved over the problem 24 hours a day for a mere six days—is the first step toward developing a diagnostic test for the virus and possibly a vaccine.

McNeil stirred in a tool story element, citing the rapid work of the sequencers. But here the application is what mattered most, since the genetic code was critical to understand what sort of enemy SARS really was. Today, rapid-fire genome sleuthing is quickly becoming a core technique for scientists responding to disease outbreaks and, as such, a recurring topic for journalists covering public health.

But good journalists go beyond just responding to a news event. Consider the approach of *New Scientist*'s Debora MacKenzie. Reading between the lines of a paper that TIGR researchers published in *Science* in 2002, she showed that the killer anthrax was most closely related to supplies held at the Army germ lab at Fort Detrick, Maryland. And when avian influenza hit Thailand in January 2004, she was soon reporting how the flu genes were closely related to samples isolated previously in Chinese duck meat.

Given her numerous scoops, I began to wonder if MacKenzie was using scientific tools herself. I had a vision of a reporter scouring gene databases like the NIH's GenBank via the Internet from her newsroom cubicle. I e-mailed MacKenzie, and she told me she leaves the science to the experts. "We're journalists," she wrote back. "We don't do the science, we report it. But we know enough about it to dig out the results that should be publicized."

I wonder. Given the pace of genomics and the volumes of genetic information now in online databases, I think it won't be long before some enterprising reporter begins using the "tools" of biology to scoop even the scientists.

29

Human Cloning and Stem Cells

STEPHEN S. HALL

Stephen S. Hall is the author of four critically acclaimed contemporary histories of science: *Invisible Frontiers* (1987) (on the birth of biotechnology), *Mapping the Next Millennium* (1992) (on new forms of cartography), *A Commotion in the Blood* (1997) (on the immune system and cancer), and *Merchants of Immortality* (2003), a winner of the NASW Science-in-Society Award, which chronicles the history of regenerative medicine, including the science, business, politics, and ethics of stem cell and cloning research. He has served as an editor and contributing writer at the *New York Times Magazine* and has also written for the *Atlantic Monthly, Science, Technology Review, Discover, Smithsonian,* and many other magazines. He lives in Brooklyn, New York, with his wife and two children, and is currently at work on a book about size, stature, and human growth.

Of the countless interviews I have conducted with scientists over the years, only once has a question prompted something of a striptease. In December of 1999, I found myself in the elegant parlor of the Union Club in New York City, chatting with a biologist named Leonard Hayflick. Although hardly a household name to the general public, Hayflick is that rare scientist whose name is permanently attached to a biological phenomenon. It is known as the "Hayflick limit," and it derives from experiments he did in the late 1950s and early 1960s showing that human cells grown in Petri dishes will predictably replicate for a certain number of cell divisions, but then hit a wall and stop dividing. This has obvious implications for cell biology, aging, and immortality (of the *in vitro* sort), and indeed the Hayflick limit has been the seed around which a spirited biological debate about the biology of aging has swirled, without definitive resolution, for about four decades now.

Because of this history, Hayflick has closely followed the recent work on the biology of aging and regenerative medicine, which in turn has made him a front-row spectator in the more recent controversies involving human embryonic stem cell research and "therapeutic cloning." At the time of my conversation with Hayflick, his longtime friend Michael West was attempting to obtain human embryonic stem cells through cloning—in a particularly controversial way, by putting human cells into egg cells from . . . cows. Almost as an aside, I asked Hayflick what he thought about West's experiments.

Hayflick replied by rolling up his pants leg. He bared enough skin to be able to point out a tiny dimple on his right knee. "The human cells he's using for the cow work came from here," he said. I had to stand up and lean over to see it, but there was undeniably a tiny divot in Hayflick's skin. The implications were stunning: Leonard Hayflick, the father of cellular senescence and one of the elder statesmen of gerontology, was allowing himself, in a manner of speaking, to be cloned. In addition to making the obvious point that even the most innocuous question can elicit a startling answer, Hayflick's reply offered another lesson, too: that colorful characters can provide a narrative thread for bringing a controversy to life.

Like any other kind of story, scientific controversies and ethical dilemmas are best conveyed by identifying, and dramatizing, crucial decision points reached by the protagonists. As a general principle for reporting any long-format piece, I always try to block out a chronology, figure out the main protagonists, and identify critical junctures in the narrative where one of the key players—it could be a doctor or scientist, but also a patient or ethicist—has to make a decision on a course of action that reveals intent in an unambiguous fashion. This moment of decision crystallizes the controversy in terms of its potential gain versus its personal, corporate, or social cost. In this case, Michael West had assumed the role of social, scientific, and ethical provocateur in the national debate over embryonic stem cell research, and his decision to pursue the cow–human (and, later, purely human) cloning experiments would have scientific, social, political, ethical, financial, and personal consequences. It provided an excellent opportunity for storytelling. Hayflick's involvement, in turn, provided an unexpected bonus: the opportunity to explore the deeper scientific history, corporate interests, and personal rivalries that shaped the present-day debate.

■ ■ ■

Controversy is always a difficult beat to cover, especially when the issues not only touch upon the quality and implications of science (can stem cells truly produce medical benefits?) but also tricky ethical issues. Is it ethically defensible to destroy a human embryo in order to harvest stem cells? Can a moral distinction be made between reproductive cloning, with its intent to produce a human child, and therapeutic cloning, with its intent to relieve human suffering?

These controversies are complicated by entangling layers of political rhetoric and hidden agendas. Companies have an agenda for advancing their corporate interests and, in the area of stem cells and cloning, have a well-documented history of attempting to mislead journalists. I was probably misled myself, even in the midst of my eagerness to use my Hayflick anecdote in my article about Michael West. When I wrote what turned out to be a cover story for the *New York Times Magazine*, I was giving free publicity to a scientist with financial and social motives for stirring the pot of controversy. In fact, a number of other publications—*U.S. News & World Report*, the *Atlantic Monthly*, *Scientific American*, and *Wired*, to name a few—all subsequently claimed various degrees of "exclusivity" in reporting the same basic story about West and his company. In some instances, these "exclusives" divulged scientific results ahead of publication in the scientific literature, or celebrated "breakthroughs" that have stirred strong revulsion in the public and among lawmakers—even while arousing considerable skepticism in the scientific community.

Which brings me to perhaps the most important journalistic asset in reporting upon a scientific controversy: critical judgment.

One of the first tasks in reporting on a controversy is to critically examine its basic assumptions. In this case: Is human cloning even feasible? Ever since the cloning of Dolly the sheep was announced in 1997, commentators have suggested that the cloning of a human child is "inevitable," that the technique itself is easy, that it is only a matter of time before someone does it. But are these assumptions really true?

First of all, one must define what a "successful" human experiment would even look like. While it is true that slamming a piece of adult DNA into an egg from which the chromosomes have been removed is technically trivial, it seems to me that the technique must reliably (indeed, without exception) produce a genetically intact, healthy individual in order to be considered safe and effective—that is, successful. What's the likelihood of that? Not very likely at all, as I began to discover. But I am a science journalist, not a scientist—so developing the tools with which to make this critical judgment was essential.

Then there's the question of whom you turn to in your search for information and good quotes. As in all aspects of science journalism, writing about stem cells and cloning requires balancing the intrigue and appeal of interesting scientific protagonists with the annoyingly necessary job of assessing whether the science they are promoting is any good. So whom do you quote: the deliciously articulate source whose science may be suspect, or the excellent scientist whose every utterance reeks of jargon, qualification, and hideous grammar? I always try to opt for good science, but that means you have to work even harder to make it intelligible. And the anecdotes you rely on have to be not just good stories, but stories that reflect a larger, rigorous reality.

Talking to experts is helpful, of course, but much more useful is hearing experts talk to each other. That's why I like to go to scientific meetings, and why I always read the short letters in journals that take exception to the data or conclusions of a published article.

In the case of human cloning, two meetings played a major role in shaping my thinking about the prospects of human reproductive cloning. In December 2001, at a meeting on regenerative medicine, biologist Tanya Dominko presented a fabulously sobering lecture on her failed attempts to clone a primate while working at the Oregon Health Sciences University in Portland. Slide after slide of hideous, misshapen, genetically doomed monkey embryos flashed on the screen while Dominko recited the "horror show" of well over 400 attempts to clone a monkey. No one sitting through that presentation could conclude that human cloning was easy, or inevitable. (Gina Kolata of the *New York Times* was, I believe, the only journalist to describe this presentation in the daily press, and yet it was of profound importance to the scientific platform on which a raging ethical debate was being played out.)

What about Dolly and all the other nonprimates that have been "successfully" cloned? Many of cloning's practitioners have argued that these animals are normal and healthy. In August 2001, however, the National Academy of Sciences held a workshop on cloning in Washington that made it clear that nuclear transfer techniques—in addition to being incredibly inefficient—almost always introduce genetic irregularities, usually in imprinted genes, into the cloned embryo. As a result, almost all the animals—including, perhaps, Dolly, who was diagnosed with premature arthritis before dying in 2003—suffer genetic anomalies that lead to either spontaneous abortion during gestation or severe developmental flaws postnatally.

Prior experience can inform your judgment about these claims. For an earlier book on cancer immunotherapy, *A Commotion in the Blood*, I had followed experiments on possible cancer drugs in mice, and I noticed that researchers often claimed that the animals were "perfectly normal," or words to that effect, when subjected to these highly toxic compounds. Those drugs often caused serious side effects when tested in humans. So I've developed a facetious rule about animal experiments: Until we can personally interview a mouse (or any other animal) to ask how it feels, it's really not very scientific to conclude that the animals are fine. I think that's an excellent rule of thumb for all animal experimentation, including cloning.

All this attention to science in a discussion of ethical controversy is not an accident. As I noted in my book about stem cell research, *Merchants of Immortality*, if it is true that the ethics of stem cell and cloning research are too important to be left to the scientists, it is equally true that the science is too important to be left to bioethicists, politicians, and other nonexperts. In any controversy involving science, you've got to get the science right before giving voice to the disputants.

■ ■ ■

Disagreement, of course, is what creates a controversy, and both sides need to be heard. It's essential to convey what the disagreement is about, but just as important not to become merely a vehicle for mischievous rhetoric. Hence, here's another somewhat heretical observation: "Balanced" reporting about a controversy is often overrated.

During the public debate leading up to President George W. Bush's stem cell decision in August 2001, there was ostensibly a "scientific" debate about the relative merits of embryonic stem cells (which require destruction of an embryo) and adult stem cells (which don't). The scientific argument was that if adult stem cells were as versatile as some argued, there was no need for the ethically vexing creation of embryonic cells. The problem with this "debate," in my opinion, was that virtually no scientists, including adult stem cell biologists, made this argument. Scientists overwhelmingly took the position that it was too early to choose one technical approach over the other. I was hard-pressed to find a single top-ranking scientist who didn't believe both adult and embryonic stem cell research needed to be aggressively pursued.

So how did the notion emerge that there was scientific disagreement on this very important point? In press reports, on televised debates, and in congressional hearings, the public repeatedly encountered a young man named David Prentice, who was billed (on his own website) as an expert on stem cell research as well as an ad hoc adviser to Senator Sam Brownback, perhaps the leading Senate opponent of embryonic stem cell research and cloning. It turns out that Prentice, then a biology professor at Indiana State University with views aligned with right-to-life groups, had no laboratory expertise in stem cell research. He had not a single peer-reviewed publication in the field, and he even admitted to me in an interview that a grant application he submitted to the National Institutes of Health to conduct research in the area had not qualified for funding. He repeatedly cited decades-old research on bone marrow transplantation as evidence that adult stem cell research was already sufficiently advanced to obviate the need for embryonic stem cells.

Bioethicist Thomas H. Murray, president of the Hastings Center in New York, has shrewdly observed that one of the mechanisms of contemporary political warfare involving scientific issues is to create the illusion of disagreement within the scientific community. "Taking a cue from the tobacco industry, pro-life operatives learned that you do not need masses of scientists on your side," Murray wrote in a 2001 essay.

> For decades tobacco lobbyists trotted out a handful of scientists who were willing to express their doubts about one or another facet of the scientific evidence linking smoking with illness and death. . . . Opponents of embryonic stem cell research hoped to impress policy makers and influence reporting by having even one scientist to provide "balance," much the way the tobacco industry salted hearings and occasionally the scientific literature with their smattering of scientific allies.

So, odd as it sounds, attempts at balance and fairness, without analysis and context, may paradoxically be misleading and mischievous.

■ ■ ■

This is what happened in the coverage of the National Academy of Sciences workshop on cloning in August 2001. From the admittedly well-intentioned impulse to be balanced, many journalists covering the meeting were, in my opinion, too scientifically kind toward several researchers who defiantly declared their intention to clone humans, despite the misgivings of scientists and society at large. These researchers—the Italian in vitro fertilization expert Severino Antinori; the Kentucky "sperm expert" Panos Zavos; and Brigitte Boisselier, scientific director of Clonaid, the cloning company affiliated with the Raelian cult—were dutifully described as "mavericks" or "rogues." But those adjectives, even in code, address attitude, not technical competence. The public—including many in Congress and the White House—undoubtedly read accounts of that meeting and reasonably concluded that the "maverick" cloners were going to be successful if they weren't stopped immediately by legislative sanctions or other blocks. The real point, rarely made, was that the bulk of evidence suggested that they couldn't succeed for scientific reasons—evidence that has only become more persuasive in the past few years.

Granted, the average length of a newspaper article typically doesn't allow one to discuss, in any degree of detail, the technical hurdles to successful human cloning. But story length in itself becomes part of the problem in discussing a lot of scientific controversies: Brevity and balance, paradoxically, can create a misleading impression of probability. The scenario reached its sorry and sordid apex in December 2002, when the Raelians claimed to have cloned a human baby named Eve. Every paper in the country felt obliged to cover the claim, which is now widely believed to have been a hoax. If the entire debate had been more grounded in science, if editors were more conversant in science (or more trusting of their reporters who *do* understand it), I believe no one besides the *National Inquirer* would have even bothered to write up the Raelian press conference.

The issue of length brings up one final bias of mine: The length of the story you're doing will dictate how much context and nuance you'll be able to convey. In brief accounts of the stem cell and cloning controversy, one has to revert to a kind of political shorthand to convey aspects of the disagreements, like the often-inflated scientific claims of potential benefit or the often-inflamed language of harm offered by the right-to-life movement. These boilerplate characterizations often come across as caricature. In the longer magazine format, I was able to recount some of the early history of the debate over embryo research in the United States, providing better historical context. In my book, I had the even greater luxury of recounting at length battles between right-to-life forces and scientists interested in human embryo research back in the 1970s, and how that connects in a very straight line with the stem cell and cloning disputes. All that early history prefigures the more recent debate. Indeed, it makes clear that what we think of, judging from our newspapers, as contemporary scientific controversies often are not very new at all.

Part Five

Covering Stories in the Physical and Environmental Sciences

In his beautiful essay on nature writing—and nature writing, by definition, *should* be beautiful—McKay Jenkins in chapter 33 makes this point about his particular style of science writing: "It is imagination or perspective or a 'way of seeing.' . . . The trick is . . . to examine the space between what we can see and what we can imagine . . ."

If it sounds exalted, well, many specialists regard their job in just that way. And having, in the introduction to part IV, praised the role of the generalist, I want here to acknowledge the importance of science writers who work in, and expand our knowledge of, specific niches in the world of science.

As anyone who has been a utility newspaper science writer knows, "science" is the broadest of all beats. Government reporters may cover City Hall. Education reporters may write about schools and school boards. Science writers may report on asteroids one day, HIV vaccine experiments the next, sonar technology the next, a universe without boundaries.

In other words, general science writing is rarely dull and rarely specific. All-purpose science writers justly take pride in their work, for the reasons I outlined in the part IV introduction and because it takes real talent and skill to keep up with the ever-changing landscape involved. There's an excellent example of that in part V, with Glennda Chui, of the *San Jose Mercury News*, discussing coverage of earth sciences (chapter 34), which happens to be one of the many aspects of science that she covers.

But specialty writing stands out in different ways. Specialized science writers possess a more finely tuned knowledge of a particular subject, and a bank of first-class sources who know and respect the writer's work. They learn to rapidly recognize innovative research from recycled ideas. They are often given more space to discuss a scientist or a trend or just a fascinating piece of research. As a result, specialized writers also gain the career-enhancing effect of becoming nationally recognized in the fields they cover.

Some journalists, myself included, move slowly in this direction. It took me 10 years of being a utility science writer before I decided that I wanted to narrow my focus, that the stories that most consistently intrigued and excited me were all based in behavioral science. Others begin as specialists, such as Ken Chang, who leads off part V, and whose graduate work was in physics and technology, two areas he now covers for the *New York Times.*

Ken offers a relaxed perspective on how much science education a good science writer needs: "You don't need to be an electrical engineer. You don't need to know how to calculate electrical impedance. (Heck, I can barely convert metric to English.) But when writing about the electrical grid, it is useful to know the difference between direct current and alternating current."

The important thing, he advises, is either being trained or teaching oneself to "learn the processes underlying what will actually make it into the article." He refers, in this instance, to the specific processes by which we can create and distribute electrical power.

That leads me very nicely to another point I want to raise both about specialists and about process—in an entirely different sense.

Science itself, of course, is a process, a work in progress, a long series of experiments on a particular question, turning up positive and negative and null results in pursuit of specific answers or with the lofty goal of testing a particular theory.

To the event-oriented news media, such results tend to be covered, logically, as events. Often they are described as a "breakthrough," which some of us have come to call the dreaded B-word. Breakthrough stories understandably get better play than process stories; they have the necessary theater and immediacy. The very word "process" tends to make news editors nod off at their desks.

But B-word stories also can seriously misstate the actual value of an experiment and a sequence of them (salt is bad for you; no, it's good; no, it's bad) serves neither readers nor the science (which, in this case, continues to suggest some very real risks linked to sodium consumption).

A writer who spends years tracking a particular science, obsessing over its details, attending its meetings, reading the insider journals, is well positioned to evaluate an extravagant new claim. Conversely, he or she may also be able to see real potential in the unheralded poster session that many reporters would walk right by.

That doesn't make such journalists into scientists, but it does make them extraordinarily able at explaining and interpreting what scientists do.

Thus, we have Usha McFarling's thoughtful discussion of covering climate change, with its knowledge of history, politics, and science, which includes an illuminating example of what can go wrong when a reporter jumps to a story without thoroughly exploring its background (chapter 35); Mike Lemonick's perspective on recognizing trends glittering at the edges of the universe before the astronomers actually report them (chapter 31); and recommendations from Andy Revkin—one of the best environmental writers working today—on how to report on an ever-changing, uncertain, and sometimes unnerving area of science (chapter 32). Cristine Russell adds to the mix her very intelligent analysis of the challenges of risk reporting (chapter 36).

Which brings me back to the idea proposed at the start of this introduction, that of a writer's ability to explore the open territory between the known and unknown. These are journalists who work constantly and well in that uncharted terrain. At a pragmatic level, that ability rests on all the hard work described above and illustrated in each of these chapters.

Science at its best works in precisely this uncharted region—extending the reach of knowledge by its cautious steps forward, its backward stumbles, the whole uncertain bumbling and amazing process of the explorer working without a map.

And science writers, at their best, learn how also to step out into unexplored realms, not only to describe the destination but to chronicle, in all its wonderful detail, the journey itself. Let us be prepared and equipped to give all such expeditions their full, unexpected and sometimes intoxicating due.

DEBORAH BLUM

30

Technology and Engineering

KENNETH CHANG

While in graduate school studying physics at the University of Illinois at Urbana-Champaign, Kenneth Chang worked at the National Center for Supercomputing Applications, which created Mosaic, the first Web browser to display Web pages containing pictures. With NCSA playing a lead role in the Internet explosion, he wrote most of the early versions of *A Beginner's Guide to HTML* and designed the center's first home page in 1993. The following year, Ken abandoned his physics work—a futile attempt to control chaos—and instead of seeking fame and fortune on the Web, he became a science writer. He attended the University of California, Santa Cruz science writing program and proceeded through a series of internships and temporary jobs before landing at ABCNEWS.com during the height of Web frenzy—too late to cash in on stock options. He joined the *New York Times* in 2000 as a science reporter covering the physical sciences.

When my home state of New Jersey deregulated its power utilities several years ago, my dad said, "Huh? How can I buy electricity from another company? It's all still the same power lines, so how can another power company send electricity to my house?"

I said, "I don't know."

Maybe you know better than I did, and the answer is obvious. Or maybe you don't know either. In either case, think about how you would try to explain this to someone else. I'll come back to the electrical grid in a bit.

This chapter is not about covering the technology beat. I've never done that. I don't write about Intel's latest chips or the iPod, and I can't tell you how to tell a bona fide hot biotech from one spewing hot air. I'm a newspapering science writer who gets to loll around in the stratified airs of Wonder and Joy of Human Knowledge much of the time, writing about galaxies or neutrinos or dinosaurs. I gleefully tell PR people pitching some techy gizmo, "I'm sorry. That sounds way too useful for me to write about."

But, of course, there is an important technology side to science writing, explaining how the science of stem cells, superconductors, nanotechnology, and so on, will impinge on everyday life—the "what does this mean to the person on the street" angle.

This part of an article can be done perfunctorily and badly. Example: Almost every article written about high-temperature superconductors—materials that can carry current with virtually no electrical resistance at relatively warm temperatures—includes the phrase "could one day lead to levitating trains."

That sounds cool. In a superficial way, it answers the question, "What are the practical applications?" But it provides little context about what scientists and engineers find intriguing about these materials, what their advantages and disadvantages are, what the hurdles are for making useful devices out of them.

Plus, high-temperature superconductors were discovered in 1986. Have you seen Amtrak levitate recently? (And the one commercial magnetically levitating train in the world, in Shanghai, does not use high-temperature superconductors.) Thus, in this case, the reporter has fulfilled the obligation of offering a potential application, but that potential is so misleading that it is a disservice to the reader.

When done right, the technological applications can also be the part of the article that brings the wonder home.

The everyday world is full of black boxes—technological machinery that for many people fully satisfies Arthur C. Clarke's dictum about being indistinguishable from magic. A Pentium computer chip consists of tens of millions of transistors. What's a transistor, and what does it do? Ask your friends and family. Ask a tech writer, even. You'll get an "Umm" or an "I don't know" from most.

Given that, what's someone likely to make of reports of the world's smallest transistor or the fastest transistor?

At one level, you don't need to know what a transistor is to know that smaller, faster ones will make faster computers. But this is the opportunity of good science writing, to pull back the curtain and let the reader in on the technological wizardry.

When I write about a smaller, better, faster transistor, I try to include explanatory sentences that say a transistor is a voltage-controlled switch, that the "on" and "off" positions of the switches represent the 1's and 0's that a computer uses to calculate. I try to explain how the switch works: Electrical current flows through a transistor like water through a garden hose, and applying a voltage is like stepping on the hose, turning the flow off.

That won't turn a newspaper reader into an electrical engineer, but it may dispel some of the mystery of technology.

Here's some advice on writing about the technological implications of science, or at least what I try to do.

Try to Oversimplify

There's the well-known Albert Einstein quote: "Everything should be made as simple as possible, but not simpler." The way to make something as simple as possible is to oversimplify—and then pull back.

Bounce your ideas off the experts. They'll correct misconceptions and help you hone the explanation.

For a *New York Times* section commemorating the 100th anniversary of the Wright Brothers flight, I wrote an article about how people still argue vehemently about how to explain how wings work. (The underlying science is all solved; it's just complicated to explain.)

One camp argues that the best explanation is simply the good ol' Newton's laws of motion, in particular the bit about "For every action, there is an equal and opposite reaction." In this explanation, a wing is simply a device that diverts air downward. By Newton's laws, the downward moving air results in the equal and opposite force pushing the wing upward. Ergo, plane flies.

I tried to oversimplify. I asked, "So it's basically air molecules bouncing off the bottom of the wing pushing it up?"

My expert said, "No."

He explained that air bouncing off the bottom of the wing does generate some lift, but most of the force actually comes from air along the top of the wing being pulled downward. He saved me from saying something that sounds completely reasonable but isn't true—the most treacherous pitfall in this business—and pointed me to the next part of the science that needed explaining. (So why does the top of the wing pull the air downward?)

Look for Everyday Units

A quick aside: If you write for American publications, find a Web Site or a computer program that converts metric units to silly, obsolete English units. It'll cut down on stupid mistakes. Better than inches or pounds or gallons, however, is finding a comparison to a familiar object. When the Lunar Prospector spacecraft discovered water on the moon, I calculated how many swimming pools it would fill.

Fish for Analogies

Corollary: Explain through concrete visual words, if possible. For example, one approach for scanning for anthrax or botulism is to use antibodies designed to attach, in a key-in-lock fashion, to proteins in those germs. Fluorescent molecules are attached to the antibodies so that they light up like bicycle reflectors when the germs are detected.

If a scientist is not giving a clear explanation, ask straight out, "Is there some sort of analogy I could use?"

A Little Expertise Is a Good Thing

You don't need to be an electrical engineer. You don't need to know how to calculate electrical impedance. (Heck, I can barely convert metric to English.) But when writing about the electrical grid, it is useful to know the difference between direct current and alternating current. Even better is to have a sense of what engineers mean when they say "in phase" or "out of phase." You don't need to go to school for this. You cannot possibly know everything that would be good to know. Rather, when reporting the story, learn about the processes underlying what will actually make it into the article.

Draw on Experience

In an article about a new technology from IBM that stores information by poking minuscule holes in a thin film, I thought, "Wow, just like the old computer punch cards." I used that, because it gave a tangible comparison point for many readers, even though I realized few people under 40 have seen punch cards—well, maybe they have seen some at the Smithsonian. The punch card analogy also illustrated a point that I was trying to make in the story, that after decades of converting from mechanical to electronic systems, now technology might actually be swinging a bit back to the mechanical.

Provide Context

So I hate "could one day lead to levitating trains" in articles about high-temperature superconductors. Then what would I prefer? Actually, I wouldn't have minded that sentence in the early articles in the 1980s if it had been followed by "But first, scientists will have to figure out how to make flexible wires out of these materials, which are brittle ceramics. High-temperature superconductors will likely first find application in high-precision scientific equipment."

Now you're not as surprised that there aren't levitating trains.

(So that you aren't misled: Scientists have since figured out how to make wires out of high-temperature superconductors, and they may indeed soon find wide use in cables for carrying large amounts of electricity. High-precision scientific equipment was indeed the first application of high-temperature superconductors. The larger reason why there aren't levitating trains is that they're expensive to build, and most people would rather just fly.)

Back to the electrical grid.

When a good chunk of the northeastern United States and eastern Canada went dark in August 2003, the largest blackout in North American history, I still had no idea how electricity is routed through the power grid, how the electricity Enron sold to California actually got to California. I blindly accepted that somehow it did.

Then after the blackout, an editor told me to write an article about technological improvements to make the grid more reliable. That meant I finally had to figure out how the thing worked.

What was confusing me was my physics education, which told me there is no way to raise or lower the current along just one branch of a circuit, no way to send electricity from point A to point B.

Robert Lasseter, a professor of electrical engineering at the University of Wisconsin, finally clued me in. I was thinking about it wrong. When my dad or California buys electricity, the electricity is not bundled into a nice parcel and shipped along the transmission lines to the destination.

It's more like a water reservoir with lots of people drinking through straws. Water flowing in from a stream (the equivalent of a power plant in this analogy) can't be directed to any particular straw. Rather, when you buy electricity, it's like paying someone to replenish the water you've drunk so that the reservoir remains full.

This light-is-finally-on insight still had to be transformed into an article. My first attempt at the top of the story was pedantic:

> In one sense, power grids are simple. The amount of current flowing through a wire is basically the voltage applied

> divided by the wire's reluctance to letting the current through, and even complex networks can be solved using a set of basic principles known as Kirchoff's Laws that are taught in high school and college physics classes.
>
> In just about every other sense, power grids are very complex.

The second try, following a suggestion by the editor, Joe Sexton, played up the reservoir analogy:

> Electricity goes where it wants to go. It follows the paths of least resistance, like water flowing downhill.
>
> That fact has perhaps been underappreciated in the past week as people struggle to understand how a blackout can sweep across nine states and a Canadian province in seconds.
>
> The power grid is not a programmable network like the phone system or the Internet where data can be sent along a specific path to a specific destination. Rather, the power system can almost be thought of as a reservoir, with power plants as streams feeding into it, and electricity consumers imbibing via straws.

Joe changed his mind and wanted something more dramatic. He offered this rewrite (Beware of e-mails from editors that start, "Ken, I played around with the top a little . . ."):

> Even in the 21st century, electricity remains a great, dangerous beast of a thing to tame, much less control. Humans, for all their ingenuity and technological contraptions, are still limited in how well they can dictate how electricity behaves—how to direct it, how to stop it.
>
> The power grid is not a programmable network like the phone system or the Internet where data can be sent along a specific path to a specific destination. And electricity is much harder to control than water, say, in a network of pipes. Electricity will not sit in a reservoir, but rather shuttles back and forth through a complex network of power lines. So the amount of power being produced at any moment must match the amount being consumed, or the system will summarily destroy itself.
>
> Those doses of humility, many experts believe, can be useful in understanding what the American public, in the 12

> angry and confused days since the biggest blackout ever, has been somewhat unknowingly referring to as the nation's grid.
>
> What the grid is, many experts agree, even if they do it quietly, is a great daily crapshoot—a slightly arrogant exercise in figuring out how millions of miles of old copper wires can light America while, at any given moment, not be overrun by the beast.

I shot back an e-mail that this was overwrought histrionics, that the grid is not continually balanced on a knife edge poised to destroy itself at any moment, and that the rarity of major blackouts shows the system works very well most of the time.

A couple more back-and-forths—retaining some of the drama, adding the level-headed perspective, weeding out the overly technical details—and we ended up with something I think we both liked. The article ran on the front page:

> The day-to-day operation of the nation's power grid is, in many respects, a great marvel—a second-by-second balancing act of the tremendously volatile thing known as electricity, a sometimes wicked creature with a mind of its own that can cause great damage in a hurry.
>
> The grid, much misunderstood, is not a programmable network like the phone system or the Internet. Electricity cannot be sent from here to there in nice packages.
>
> Rather, the grid is like a giant invisible reservoir where the amount of power being put in at any moment must match the amount being consumed.

A few days later, Victor Mather, a copy editor on the national desk, someone I didn't even know well, came up and said that of the mountains of articles that ran after the blackout, mine was the first that made sense to him in explaining the workings of the grid.

That was good to hear. The article had turned out to be more than just an exercise in answering my dad's questions years too late. I felt that I had done something useful, too.

31

Space Science

MICHAEL D. LEMONICK

Michael D. Lemonick is a senior writer at *Time,* where he has worked since 1986. He has written more than 40 cover stories, on a wide variety of scientific topics, including supernovas, superconductivity, particle physics, Egyptology, and cosmology, and has twice won the American Association for the Advancement of Science-Westinghouse Award for science writing. Mike graduated from Harvard College in 1976 with a bachelor's degree in economics, and received a master's degree in journalism from Columbia University in 1983. He began his career at *Science Digest* magazine and served briefly as executive editor of *Discover* magazine. He has written freelance pieces for *Discover* (six cover stories), *People, Science 83, American Health, Audubon, Playboy,* and the *Washington Post.* He is also the author of three books on astronomy: *The Light at the Edge of the Universe* (1993); *Other Worlds* (which was awarded the 1998 American Institute of Physics Science Writing Award); and *Echo of the Big Bang.* (2003).

Astronomy is the only branch of science where the questions are literally cosmic. Its practitioners are trying to answer the most profound questions imaginable, the same questions that philosophers have been wrestling with for thousands of years. How big is the universe? How old is it? What is it made of? Are we alone, or do other intelligent beings live on planets orbiting distant stars? How did the cosmos begin? And how will it end?

As recently as a decade ago, none of these questions had been answered in any definitive way. Now, thanks to powerful new space-based observatories and ingenious new techniques for gazing up from the ground, astronomers have cracked some of them. We now know that the universe is 13.7 billion years old, that more than 100 planets circle Sun-like stars right in our celestial neighborhood, and that the cosmos is likely to expand forever, until all the stars have burned out and matter itself breaks down. We know that gamma ray bursts—explosions so massive they defied understanding for decades—are exploding stars more powerful than anyone had imagined.

Yet plenty of mysteries remain. Astronomers know that the visible stars and galaxies add up to only a fifth or so of the matter in the universe. The rest is some sort of mysterious dark matter, detectable only through its gravitational influence on the visible stuff. The search for dark matter is still a major focus of modern astronomy. Closer to home, there's a major push to find not just planets, but Earthlike planets orbiting nearby stars. The massive, gaseous, Jupiter-like planets found so far are impressive enough, but as far as we know, you need something smaller and more solid to support life—the ultimate goal of planet-searchers. Indeed, while astronomers had long since given up looking for life in our own solar system, biologists have given them new hope. Life, it turns out, can live in far harsher conditions than anyone thought (hot springs, Antarctic ice, inside solid rock), which means it could exist under the surface of Mars or in oceans under the icy coating of Jupiter's moon Europa.

And there are always surprises. In 1998, astronomers found evidence for something called dark energy, long considered to be a purely theoretical notion. It turns out that the universe seems to be pervaded with a bizarre antigravity force that's pushing galaxies apart faster and faster all the time. Nobody has a clue what the source of this astonishing energy might be. Astronomers are wrestling with these and dozens of other mysteries every day at observatories and in front of computer screens, and meeting with each other all the time to discuss (and frequently fight over) the latest evidence.

By now it may be possible for the discerning reader to guess that I have more than a passing interest in astronomy. When I became an astronomy writer, the challenge was to make everyone else—those who aren't passionate about the inflationary universe—share that fascination. I had to understand the science deeply enough to be able to restate it in my own words, clearly and accurately. Concepts like cold dark matter, nuclear fusion, the curvature of the universe, and angular momentum, which I'd always just accepted, now had to be part of my working vocabulary. I'd always read about astronomy for fun. Writing about it was like learning a foreign language after getting along in a pidgin version for years.

And just as is the case with learning a language, the only way to absorb these ideas was through constant repetition. I need to hear something two or three times (and sometimes 20 or 30 times) before I truly, intuitively grasp it. So I read everything I can get my hands on: newspapers with good astronomy writers (*New York Times, Washington Post, Boston Globe, Los Angeles Times, San Francisco Chronicle, San Jose Mercury News, Dallas Morning News*); popular magazines like *Discover*, specialty astronomy magazines like *Sky & Telescope* and *Astronomy*, and the more technical journals, including *Science, Nature*, and the *Astrophysical Journal*.

I also read plenty of astronomy-related books. I urge people to read Marcia Bartusiak's *Through a Universe Darkly* (1993), Dennis Overbye's *Lonely Hearts of the Cosmos* (1991), and Alan Dressler's *Voyage to the Great Attractor* (1994). Stay away from Stephen Hawking's *A Brief History of Time* (1988), which was falsely advertised as being comprehensible. Instead, try his 2002 book *The Universe in a Nutshell*—a wonderful, informative, and beautifully illustrated volume. Modesty forbids me to suggest one of my own books. But if you insist, I won't object.

Books are for background knowledge. To find out what's going on right now, I rely largely on informants who help me wade through the thousands of technical papers, talks, and conference reports that come out each year. Among these helpful folks are university public relations officers and editors at the major broad-spectrum science journals. Both *Science* and *Nature*, for example, send an extra signal, a sort of raised editorial eyebrow, when they think an article is especially interesting, by publishing an accompanying news story. (It's usually more comprehensible than the original paper.) They also let journalists know a week in advance, confidentially, what papers they'll be publishing, giving us a leg up on being ready with a thoughtful story when the journal actually appears.

Astronomy itself, moreover, in the form of its professional society, the American Astronomical Society, has had an unusually effective press officer. Steve Maran, an astronomer himself at NASA's Goddard Space Flight Center and also a talented science writer, is among the best in the business at spotting major stories and alerting reporters about them. He stages press conferences at the association's two big meetings each year, puts reporters in touch with astronomers, and helps explain a report's significance.

Most important of all in developing stories, though, is to cultivate relationships with astronomers themselves, so that you can find out what they're doing before it gets as far as a press release of a journal paper. How do you cultivate such relationships? It takes time, but it isn't especially complicated. The basic requirement is common sense.

Astronomers are just like anyone else; they are generally happy to try to help you (and therefore your readers) understand what a particular observation or theoretical breakthrough is and why it's important. All they ask in return is that you do your best to get it right. That involves asking questions, and being unafraid to ask dumb ones. You don't want to be too dumb if you can help it;

that's why constant reading is important. It's okay to ask what a quasar is, if you really don't know—but even a novice astronomy reporter should know.

However, most scientists don't expect you to know their fields thoroughly at first; if you don't ask questions when you don't understand something, you'll earn a reputation as someone who doesn't really care—or, what's just as bad, as someone who can't tell the difference between grasping a concept and missing it entirely. In general, an astronomer would rather that you ask a question three times to make sure you've got it than have you ask just once and print a lot of garbled nonsense.

Even so, I generally raise the possibility of a follow-up interview. Even after two decades of experience, I sometimes get back to the office, start writing my story, and discover that I've failed to ask a crucial question. If I've warned the scientist that this will undoubtedly happen, and have permission in advance for a follow-up, they're always amenable. After all, the same sort of thing happens to them. Since I may find myself baffled after working hours, I always get both an office and a home phone number (and permission to use the latter), plus an e-mail address.

I also warn my sources in advance that I might be asking them to look at all or part of the story before it goes to press. This is still considered anathema to some reporters, and rightly so in political or investigative reporting where you may be making an accusation or uncovering some unsavory facts. But when you're explaining the evidence for a giant black hole at the core of a distant galaxy, the risk of an astronomer backing away from a story or calling a lawyer is minimal, while the risk of getting it wrong is real.

All these good intentions don't count for much if the story that appears in print isn't any good. This can happen if thorough reporting is not accompanied by clear writing. It is all too common to read a story in which all the facts are correct but the overall impression is one of garble and confusion. The rules for good science writing are the same as those for good writing in general: Be clear; structure your story so that it flows logically. The reader should reach the end of each paragraph thinking "I want to know more," not "Can I stop now?" Use too few adjectives rather than too many; that way you'll avoid sounding gushy.

Here's an example that I hope illustrates these points, taken from a 2000 cover story on the fate of the universe that I wrote for *Time.* This passage describes what happened when two independent groups of observers tried to measure how much the expanding universe had slowed down since the Big Bang, by looking at supernovas, or exploding stars.

> By 1998 both teams knew something very weird was happening. The cosmic expansion should have been slowing down a lot or a little, depending on whether it contained a lot of

> matter or a little—an effect that should have shown up as distant supernovas looking brighter than you would expect compared with closer ones. But, in fact, they were dimmer—as if the expansion was speeding up. "I kept running the numbers through the computer," recalls Adam Riess, the Space Telescope Science Institute astronomer analyzing the data from Schmidt's group, "and the answers made no sense. I was sure there was a bug in the program." Perlmutter's group, meanwhile, spent the better part of the year trying to figure out what could be producing its own crazy results.
>
> In the end, both teams adopted Sherlock Holmes' attitude: Once you have eliminated the impossible, whatever is left, no matter how improbable, has got to be true. The universe was indeed speeding up, suggesting that some sort of powerful antigravity force was at work, forcing the galaxies to fly apart even as ordinary gravity was trying to draw them together. "It helped a lot," says Riess, "that Saul's group was getting the same answer we were. When you have a strange result, you like to have company." Both groups announced their findings almost simultaneously, and the accelerating universe was named Discovery of the Year for 1998 by *Science* magazine.

It's also important to define your terms when necessary, judging from experience what "necessary" means. So, for example, you don't have to tell readers what a star or a planet is. But you might have to tell them what a galaxy is, at least in passing. And you'll probably have to define quasar, cosmic microwave background, and redshift.

Finally, you have to learn what to leave out. Just about everyone has had the excruciating experience of listening to an inept raconteur telling an endless story full of unimportant details. It's hard to extract the signal from the background noise. Science writing is analogous, though not precisely the same. There are few if any unimportant details in a piece of scientific research. But some facts and logical steps and concepts are more important than others—and if you include everything, the average reader is going to end up hopelessly confused.

Since that's the opposite of what we want to achieve, good science writers do leave out lots of the details, perhaps leaping over three steps in the scientist's deductive process in order to get to a fourth. We know we're doing it, and the scientists understand that we have to. The best science writers come out of this balancing act—between drowning the reader in facts and oversimplifying to the point of inaccuracy—with a story that is both reasonably accurate and readable.

If I have any doubts about whether I'm striking the right balance, I try to explain the story to an acquaintance who doesn't know much about science, and also to a scientist. If they both like it, I've done my job: I've written a good, accurate, newsworthy story. If I do that over and over, astronomers will know they can trust me, and they'll begin calling me, volunteering information about work in progress, about stories coming up that I might want to cover. They'll also be receptive when I call them and ask what's the most interesting or exciting thing they've heard on the astronomical grapevine lately. Almost everything worth talking about has been discussed at length among astronomers before it's made public. You might as well know what they've been discussing.

Finally, while it's very satisfying to come upon a piece of breaking news long before it breaks, it's even more satisfying to detect a broad trend well before it happens. That's where a serious basic interest in the subject—and a layperson's sense of astonishment at the mind-bending phenomena the universe conceals—comes in handy.

When I heard someone speak at an astronomical conference in 1983 about something called gravitational lensing—a cosmic optical illusion formed when a nearby galaxy's strong gravity creates a double image of a more distant quasar—the concept itself struck me as being so strangely cool that I sold my editor on a feature story. It turned out to be one of the first on what has become, two decades later, one of the hottest fields in astrophysics. Among other things, astronomers have used gravitational lensing to make precise, direct measurements of the distance to faraway galaxies, to determine the mass and distribution of dark matter in clusters of galaxies, to find planets passing in front of distant stars, and, early in 2004, to find a galaxy so distant that it must have formed just a few hundred million years after the Big Bang—and all thanks to a phenomenon Einstein once said would never actually be observed.

And that's the other thing that makes me love astronomy so much: There's always something new and very surprising going on. In 2003, astronomers using powerful infrared telescopes discovered an object they called Sedna—three times farther away than and about two-thirds the size of Pluto, and quite likely just one of many similar worlds that inhabit the outer reaches of the solar system. And even as it's allowing the Hubble Telescope to die, NASA is moving ahead with a new space observatory, called the James Webb Space Telescope, which should be able to probe deep into the Cosmic Dark Ages, when the first stars were born. And theorists are tearing their hair out trying to understand the nature of the dark energy that's causing cosmic expansion to speed up. The next few years will be a wonderful time to be writing about astronomy.

32

The Environment

ANDREW C. REVKIN

Andrew C. Revkin has written about science and the environment for two decades, most recently as a reporter for the *New York Times.* He is the author of *The Burning Season: The Murder of Chico Mendes and the Fight for the Amazon Rain Forest* (1990 and 2004). He has received many honors, including the National Academies Science Communication Award, two journalism awards from the American Association for the Advancement of Science, and an Investigative Reporters and Editors Award.

Hindsight is usually expressed in bravado-tinged phrases. "You have it so easy now" is one. But when scanning the recent history of environmental news, the impression is just the opposite. A few decades ago, anyone with a notepad or camera could have looked almost anywhere and chronicled a vivid trail of despoliation and disregard. Only a few journalists and authors, to their credit, were able to recognize a looming disaster hiding in plain sight. But at least it was in plain sight.

The challenges in covering environmental problems today are far greater, for a host of reasons. Some relate to the subtlety or complexity of most remaining pollution and ecological issues now that glaring problems have been attacked. Think of non-point-source pollution, such as runoff from countless farm fields or urban lawns, and then think of the ultimate point of the Exxon *Valdez*, spilling its heavy load of crude oil into the seas off the Alaskan coast.

A little reflection is useful. Most journalists of my generation were raised in an age of imminent calamity. Cold War "duck and cover" exercises regularly sent us to the school basement. The prospect of silent springs hung in the wind. We grew up in a landscape where environmental problems were easy to identify and describe. Depending on where you stood along the Hudson River's banks, the shores were variously coated with adhesive, dyes, paint, or other materials indicating which riverfront factory was nearest. And, of course, the entire river was a repository for human waste, making most sections unswimmable. Smokestacks were unfiltered. Gasoline was leaded.

Then things began to change. New words crept into the popular lexicon—smog, acid rain, toxic waste. At the same time, citizens gained a sense of empowerment as popular protest shortened a war. A new target was pollution. Earth Day was something newspapers wrote about with vigor, not an anachronistic, even quaint, notion. Republican administrations and bipartisan Congresses created a suite of laws aimed at restoring air and water quality and protecting wildlife. And, remarkably, those laws began to work.

Right through the 1980s the prime environmental issues of the day—and thus the news—continued to revolve around iconic incidents, mainly catastrophic in nature. First came Love Canal, with Superfund cleanup laws quickly following. Then came the horrors of Bhopal, which generated the first right-to-know laws granting communities insights into the chemicals stored and emitted by nearby businesses. Chernobyl illustrated the perils that were only hinted at by Three Mile Island. The grounding of the Exxon *Valdez* drove home the ecological risks of extracting and shipping oil in pristine places.

Now, however, the nature of environmental news is often profoundly different, making what was always a challenging subject far harder to convey appropriately to readers. By appropriately, I do not just mean accurately. Any stack of carefully checked facts can be accurate but still convey a warped sense of how important or scary or urgent a situation may be. Therein lies an added layer of responsibility—and difficulty—for the reporter.

As recently as the first days of 2004, those difficulties still made a *New York Times* colleague of mine, the veteran medical writer Gina Kolata, nearly tear her hair out as she grappled with a new paper in the journal *Science*, positing that farmed Atlantic salmon held much higher levels of PCBs and other contaminants in its flesh than did wild Pacific salmon. The authors calculated that the risks from these chemical traces meant consumers should not eat more than one salmon meal a month despite the many health benefits conferred by such fish. The Food and Drug Administration, noting that the detected concentrations were dozens of times lower than federal limits, strongly disagreed. Some top toxicologists not aligned with the seafood industry or anyone else also fervently disputed the researchers' risk calculation.

Gina carefully explained the new findings. At the last minute, to make sure hurried readers put things in perspective, she added a vital extra clause to the opening sentence: "A new study of fillets from 700 salmon, wild and farmed, finds that the farmed fish consistently have more PCBs and other contaminants, *but at levels far below the limits set by the federal government*" (emphasis added).

Even with that proviso, and a host of researchers' voices further down in the story stressing that the health benefits of eating salmon were clearer than any small risk from PCBs, readers remained confused. Salmon piled up in some supermarkets. My brother, a cardiologist and heart-drug researcher, sent me an urgent e-mail asking: "What's the poop on the risk of farmed salmon, dioxin, and PCBs? Any truth to it? I eat it as much as three times a week."

I'd covered PCBs for years in the context of the remaining stains buried in the Hudson's river-bottom mud. My own instinct on this, which I conveyed to my brother not as a journalist but simply as a citizen who has had to make judgments in the face of uncertainty, was that he should eat and enjoy (while perhaps avoiding the brownish fatty tissue and—sad to say for sushi-roll lovers—the salmon skin).

But that was instinct—or common sense.

So what is a reporter to do? The first step is simple: Know thine enemy. Recognize where the hurdles to effective environmental communication reside so you can prepare strategies to surmount or sidestep them.

Here are some of the fundamental characteristics of the news process that I feel impede or distort environmental coverage.

The Tyranny of the News Peg

News is almost always something that happened today. A war starts. An earthquake strikes. In contrast, most of the big environmental themes of this century concern phenomena that are complicated, diffuse, and poorly understood. The runoff from parking lots, gas stations, and driveways puts the equivalent of 1.5 Exxon *Valdez* loads' worth of petroleum products into coastal ecosystems each year, the National Research Council recently found. But try getting a photo of that, or finding a way to make a page-one editor understand its implications.

Here's how I handled that story, which the *Times* science editor pitched for page one, but which was trimmed back and ran on A14 on May 24, 2002:

> Most oil pollution in North American coastal waters comes not from leaking tankers or oil rigs, but rather from countless oil-streaked streets, sputtering lawn mowers and other

> dispersed sources on land, and so will be hard to prevent, a panel convened by the National Academy of Sciences says in a new report.
>
> The thousands of tiny releases, carried by streams and storm drains to the sea, are estimated to equal an Exxon *Valdez* spill—10.9 million gallons of petroleum—every eight months, the report says.

Out of all environmental stories these days, none is both as important (to scientists at least) and as invisible as global warming. Many experts say it will be the defining ecological problem in a generation or two and that actions must be taken now to avert a huge increase in heat-trapping emissions linked to warming. But you will never see a headline in a major paper reading: "Global Warming Strikes—Crops Wither, Coasts Flood, Species Vanish." All of those things may happen in coming years, but they will not be news as we know it.

Developments in environmental science are almost by nature incremental, contentious, and laden with statistical analyses including broad "error bars." In the newsrooms I know, the word "incremental" is sure death for a story, yet it is the defining characteristic of most research.

Faced with this disconnect, reporters and editors are sometimes tempted to play up the juiciest—and probably least certain—facet of some environmental development, particularly in the late afternoon as everyone in the newsroom sifts for the "front-page thought" in each story on the list. They do so at their peril, and at the risk of engendering even more cynicism and uncertainty in the minds of readers about the value of the media—especially when one month later the news shifts in a new direction. Keep watching for the tide to change on salmon and health. It will change, and change again.

Is it good enough for a story to be "right" for a day? In the newsroom the answer is mainly yes. For society as a whole, I'm not so sure.

The hardest thing sometimes is to turn off one's news instinct and insist that a story is *not* "frontable," or that it deserves 300 words and not 800. Try it some time. It violates every reportorial instinct, but it's doable—kind of like training yourself to reach for an apple when you crave a cookie.

The Tyranny of Balance

As a kind of crutch and shorthand, journalists have long relied on the age-old method of finding a yea-sayer and naysayer to frame any issue, from abortion to zoning. It is a quick, easy way for reporters to show they have no bias. But it is also an easy way, when dealing with a complicated environmental issue, to perpetuate confusion in readers' minds and simply turn them off to the idea that media serve a valuable purpose.

When this form is overused, it also inevitably tends to highlight the opinions of people at the edges of a debate instead of in the much grayer middle ground, where consensus most likely lies. I can't remember where I first heard this, but the following maxim perfectly illustrates both the convenience of this technique and its weakness: "For every Ph.D., there is an equal and opposite Ph.D."

One solution, which is not an easy one, is to try to cultivate scientists in various realms—toxicology, climatology, and whatever else might be on your beat—whose expertise and lack of investment in a particular bias are established in your own mind. They should be your go-to voices, operating as your personal guides more than as sources to quote in a story.

Another is what I call "truth in labeling." Make sure you know the motivation of the people you interview. If a scientist, besides having a Ph.D., is a senior fellow at the Marshall Institute (an industry-funded think tank opposed to many environmental regulations) or Environmental Defense (an advocacy group), then it is the journalist's responsibility to say so.

In a recent piece on climate politics, this is how I described Pat Michaels, a longtime skeptic on global warming who is supported in part by conservative or industry-backed groups:

> "Climate science is at its absolutely most political," said Dr. Patrick J. Michaels, a climatologist at the University of Virginia who, through an affiliation with the Cato Institute, a libertarian group in Washington, has criticized statements that global warming poses big dangers.

Such a voice can have a legitimate place in a story focused on policy questions, but is perhaps best avoided in a story where the only questions are about science. The same would go for a biologist working for the World Wildlife Fund.

Heat Versus Light

One of the most difficult challenges in covering the environment is finding the appropriate way to ensure a different kind of balance—between the potent heat generated by emotional content and the sometimes less compelling light of solid science and statistics.

Consider a cancer cluster. A reporter constructing a story has various puzzle pieces to connect. There is the piece brimming with the emotional power of the grief emanating from a mother who lost a child to leukemia in a suburb where industrial effluent once tainted the water. Then there is the piece laying out the cold statistical reality of epidemiology, which might in that instance never be able to determine if contamination caused the cancer.

No matter how one builds such a story, it may be impossible for the reader to come away with anything other than the conviction that contamination killed.

Prime examples of the choices journalists make in balancing "heat versus light" came amid the uncertainty and fear and politicians' assurances and activists' hype following the collapse of the World Trade Center towers. Lower Manhattan was shrouded in powdered cement, silica, plaster, and asbestos. I was immersed in the story along with a host of colleagues from media both local and long distance. Some exploited the fear and peril, drawing headlines in big block letters. Some of us tried to do something dangerous—stress the things that were not known or indeed were unknowable, even as we wrote definitively about the one risk that was crystal clear, the risk faced by unprotected workers clambering in the smoldering wreckage.

We were criticized by some media analysts for ignoring warnings that danger lay in the dust that settled outside Ground Zero. But I stand behind every word, except for one phrase written in a hurried moment (that tyranny of time!) in which I incorrectly wrote that no harmful compounds had been detected in the air. In fact they had been detected within the perimeter of Ground Zero, but at minute levels—and never outside the immediate vicinity.

Those who highlighted potential perils focused on statements by some scientists and testing companies eager for the spotlight who judged the asbestos risk around the area against thresholds devised for chronic occupational exposure—a totally different situation. Fears grew and facts were few.

Someday, perhaps two generations after 9/11, there will be sufficient time for patterns in cancer rates among exposed populations to show an effect. But anyone claiming a clear and present danger in those early days was—to my mind at least—being irresponsible.

But were they doing their job? By the metric of the media, the answer is probably yes. Pushing the limits is a reporter's duty. Finding the one element that's new and implies malfeasance is the key to getting on the front page. I'm just as attuned to that as any other reporter. All I hope for in my own work and that of others is an effort to refine purely news-driven instincts, to try to understand—and convey—the tentative nature of scientific knowledge, to retain at least some shades of gray in all that black and white.

The Great Divide

There is one way that journalists dealing with the environment can start working on building reflexes that improve that balance of heat and light, boost the ability to convey the complex without putting readers (or editors) to sleep, and otherwise attempt to break the barriers to effective communication with the public.

This is to communicate more with scientists. By getting a better feel for the breakthrough–setback rhythms of research, a reporter is less likely to forget that the state of knowledge about endocrine disruptors or PCBs or climate is in flux. This requires using those rare quiet moments between breaking-news days—sure, there aren't many—to talk to ecologists or toxicologists who aren't on the spot because their university has just issued a press release.

The more scientists and journalists talk outside the pressures of a daily news deadline, the more likely it is that the public—through the media—will appreciate what science can and cannot offer to the debate over difficult questions about how to invest scarce resources or change personal behaviors.

There is another reason to do this. Just as the public has become cynical about the value of news, many scientists have become cynical, and fearful, about journalism. Some of this is their fault, too. I was at a meeting in Irvine, California, on building better bridges between science and the public, and one researcher stood up to recount her personal "horror story" about how a reporter totally misrepresented her statements and got everything wrong. I asked her if she had called the reporter or newspaper to begin a dialogue not only on fixing those errors, but preventing future ones.

She had not. She had never even considered it.

Until the atmosphere has changed to the point where that scientist can make that call, and the reporter respond to it, everyone has a lot of work to do.

33

Nature

MCKAY JENKINS

McKay Jenkins holds degrees from Amherst, Columbia's Graduate School of Journalism, and Princeton, where he received a Ph.D. in English. A former staff writer for the *Atlanta Constitution,* he has also written for *Outside, Orion,* and many other publications. He is the author of *The Last Ridge: The Epic Story of the U.S. Army's 10th Mountain Division and the Assault on Hitler's Europe* (2003), *The White Death: Tragedy and Heroism in an Avalanche Zone* (2000), and *The South in Black and White: Race, Sex, and Literature in the 1940s* (1999) and the editor of *The Peter Matthiessen Reader* (2000). His latest book is *Bloody Falls of the Coppermine: Madness, Murder, and the Collision of Cultures in the Arctic, 1913* (2005). McKay is the Cornelius A. Tilghman Professor of English and a member of the Program in Journalism at the University of Delaware. He lives in Baltimore with his family.

Not long ago, at the beginning of a course I was teaching on "The Literature of the Land," I asked my undergraduate journalism students why they were having such a hard time thinking of things to write about. What, I wondered, was so hard about nature writing?

A sophomore raised his hand. As often happens, the answer came back more succinct than I could have hoped. "It's hard writing about nature in Delaware," he said, "because there is no nature in Delaware."

There was something emblematic in this comment, something that revealed the difficulty, at first blush, that young writers have in conjuring exactly what "nature writing" means. My first impulse was to list all the nearby "nature" out there that the student hadn't bothered to recognize: the Atlantic seashore, the Delaware and Chesapeake bays, the Appalachian Mountains on one hand; and DuPont chemical factories, massive landfills, and rampant suburban sprawl on the other. But instead I paused, and let the comment hang in the air for a moment. What, exactly, were we talking about?

For the nonspecialist, "nature writing" can seem especially intimidating, since it seems, at first glance, to be a subject without human drama, without a narrative trajectory, without a beginning, a middle, and an end—as opposed to, say, writing about cops, or courts, or politics, or sports. It can seem overly technical, or ponderous, or misanthropic. It can seem abstract, even irrelevant, especially to urban audiences who think of "nature" as something they encounter on boutique holidays out west. Norman Maclean's *A River Runs Through It,* according to legend, was rejected by a New York publisher because "it had too many trees in it."

But it isn't "nature" that is lacking, in Delaware or anywhere else. It is imagination, or perspective, or a "way of seeing." Granted, a place like Delaware is notably lacking in the 14,000-foot mountains, Arctic fjords, and equatorial rainforests that have come to represent "nature" for suburban Americans. But this is precisely why a place like Delaware turns out to be such a useful place to talk about nature writing. The trick is to see the subtleties and the synecdoches, to examine the space between what we can see and what we can imagine, to ponder the "shadow" that T. S. Eliot writes falls "between the idea/And the reality." Nature writing is more about the sharpness of the eye and the clarity of the mind than it is about the majesty of the landscape.

I say this only to separate nature writing from "environmental reporting," which tends at once to be less preoccupied with metaphysics and more with chronicling the endless tug of war in politics, economics, and environmental advocacy. The fields overlap; they both, for example, rely substantially on field research. But where environmental reporters might use this research to bolster a particular argument, a nature writer might use it as a prompt for meditation. The prospect of a manned mission to Mars provides ample opportunities for both. So does an ocean made barren by overfishing, or a plan to reintroduce the wolf. Science, in other words, can be used as an end, as an advance in an ongoing story, or it can be used as a means, to open our eyes to see larger and larger contexts.

Since so much of nature writing concerns itself with the nonhuman world, one of the struggles is to figure out how to describe and muse about things to which humans have limited access. To my mind, a nature writer has the challenge of the poet: With lofty, often abstract imaginative aspirations, he or she must find the most vivid details with which to express them. Aldo Leopold, watching the "fierce green fire" drain from the eyes of a wolf he has just killed, realizes he must stop thinking like a man and start thinking like a mountain. Rachel Carson, remembering trucks driving through mid-century suburban neighborhoods spraying lawns with DDT, makes us see not just the hazards of pesticides but the hubris of technology itself. DDT is a subject for environmental reporting. Hubris is a subject for nature writing.

Teaching this idea to my students, I often draw a diagram of a small circle with an arrow pointing to a large circle. The larger circle is the abstract idea: species extinction, global warming, the biology of death, the mind of a wolf. The smaller circle is the detail, the observation, the interview, the expedition, that gives the reader access to the larger idea. In some ways, filling the smaller circle is as hard as filling the first. Given an impulse to explore abstract ideas, how do we devise a narrative strategy to get the ideas across? How can we concoct the teaspoon of sugar to help the medicine go down? Bookshelves are full of excellent examples, any one of which can be read as models of structure and tone. The one thing most have in common, like any good piece of nonfiction writing, is a narrative arc: tales of expeditions, natural disaster, spiritual pilgrimage, ethnography, or scientific exploration that serve as a frame on which to stretch larger philosophical questions.

David Quammen says this nicely in his essay "Synecdoche and the Trout" in *Wild Thoughts from Wild Places* (1998): A trout is both a fish and an idea, a representation of something larger, in this case an entire ecosystem. The trick for the nature writer is to remember both the trout and the watershed. Write statistically about the numbers of trout living in a single stream and you miss both larger ecological implications and the metaphysics of a creature whose essence you can only approximate. Write abstractly about the health of the northern Rockies and you miss the poetic specificity of the fish. Good writing needs both.

In my own book *The White Death*, I tried, in effect, to stitch two threads together: a human narrative, about five boys killed in a mountaineering disaster in Montana's Glacier National Park; and the natural history of snow and avalanches themselves. The book moves back and forth between, on one hand, an attempt at an historic climb, a catastrophic accident, and an unprecedented search-and-rescue; and on the other, a chronicle of the deep history, science, and folklore of one of nature's most mysterious and ominous forces. In a few places in the book, I tried to combine the two, to describe an avalanche-prone mountain as a kind of stage on which human dramas have often played out. Since not all readers have been in avalanche country, I decided to use a more universal source of anxiety.

> Hiking or skiing in avalanche country is like walking around in a valley you know to be inhabited by grizzly bears.

> Your senses become more alert. You become aware of tiny sounds—every creak of a tree limb, every snap of a twig. In bear country, you become aware, perhaps for the first time in your life, that you are not at the top of the food chain. For once, nothing is so important as the direction of the wind; there is something out there that, with a mix of your own ignorance and bad luck, could finish you off. The same is true in the winter backcountry. When every footstep, on a steep slope, is potentially your last, you tend to pay attention to where you put your feet. The beauty of this arrangement is that this vibrancy, this forced concentration, makes the whole picture sharper. Time slows down. Your actions matter.

I had a different challenge in the book that followed, *The Last Ridge.* To begin with, I was drawn to the story not because I had an interest in military history but because I was interested in the veterans who returned from the war to become the country's most important mountaineers, skiers, and conservationists. As a writer who believes firmly in the balance between fieldwork and archival research—especially for a book that relied on thousands of pages of letters and military documents—I felt it was critical to see the division's training grounds and battlefields firsthand. Since this required me to travel to Colorado and Italy, this was not an onerous task. But there were a number of important aesthetic reasons for tromping around these places. Many of the young men who signed up for this experimental division had been born and raised in New England and had never been west before they arrived to train at Camp Hale, 9,200 feet up in the Colorado Rockies. As their training went on, they would spend weeks at a time living outside at 13,000 feet, even in winter. Given this, I wanted to see the place as they had, with only New England mountains in their collective mountaineering experience. To my mind, the moments of physical description would also allow the reader to breathe a bit between what are often torturous scenes of physical hardship and violence.

> When it finally arrived, spring weather also meant an explosion of color in the mountains, where wildflowers bloomed in the alpine meadows. Waterfalls poured over cliffs lining the eastern edge of camp, along the trail leading to Kokomo Pass. The air suddenly took on a hint of sage and buzzed with the dry rattle of grasshoppers. Mountain jays flitted from lightning-blackened tree snags to scraggly dead sage bushes sitting like elk antlers against the rocky peaks. In warm weather, the men could fish and play football in the mead-

> ows. But even in spring, daylight in the valley was in short supply. With so many high peaks around, the horizon was 4,000 feet higher than the men's barracks. As the aspens leafed out, their bark a waxy skin of white suede, they reminded the New Englanders of the paper birches back home. The mountains themselves, their slopes scarred by rock slides, seemed a cross between the jagged White Mountains of New Hampshire and the rounded Green Mountains of Vermont; swelling up to the east of Cooper Hill, the smooth sides of Chicago Ridge looked like a chunk of Vermont dropped from the moon.

Annie Dillard's writing is remarkable both for the vividness of the descriptions and for the arching wonder of the mysteries they evoke. Witnessing a total solar eclipse, which washed all the color out of the central Washington State hills on which she stood, she experienced a transcendent fear:

> The hole where the sun belongs is very small. A thin ring of light marked its place. There was no sound. The eyes dried, the arteries drained, the lungs hushed. There was no world. We were the world's dead people rotating and orbiting around and around, embedded in the earth's crust, while the earth rolled down. . . . The meaning of the sight overwhelmed its fascination. It obliterated meaning itself. *Teaching a Stone to Talk* (1982)

A news story about an eclipse is environmental reporting. Describing a glimpse of the end of the world is nature writing.

Bill McKibben, in *The End of Nature* (1989), writes that our addiction to fossil fuels has so damaged Earth's atmosphere that human beings have become something more than biological creatures. We have become a force of nature, like the Sun, or gravity, and McKibben does not seem confident in our ability to wield such power wisely. Barry Lopez, in *Arctic Dreams*, notes that the Inuit call Europeans "the people who change nature." Transmitting the latest research about global warming is environmental reporting. Musing on our ability to wield powers once limited to Greek gods is nature writing.

Peter Matthiessen, in his first published piece of nonfiction, imagined the moment when two fishermen smashed the last Great Auk egg ever seen by man. Another species gone for good. "Man, striving to imagine what might lie beyond the long light years of stars, beyond the universe, beyond the void, feels lost in space; confronted with the death of species, enacted on earth so many times before he came, and certain to continue when his own breed is gone, he is forced to face another void, and feels alone in time." Since Matthiessen wrote those lines, in 1959, things have gotten considerably worse. Humanity has stomped its boot heel on the world's ecology, causing rates of extinction not seen since Earth was hit by a giant meteor. We are being compared to other "weed species," except that unlike rats, crows, and cockroaches, we are actually responsible for turning lush ecosytems into wastelands. The "good news" is that Earth might recover its ecological richness 10 million years after humans themselves become extinct. This is not far from the moment in the film *The Matrix*, when the snarling Agent Smith barely has to think to come up with a creature most resembling human beings: the virus. In *The Hot Zone* (1994) Richard Preston goes a step further. To an Ebola virus, hidden inside a human host about to land at a New York airport, Manhattan looks like a meat locker. Human beings are not at the top of the food chain after all. Nature writing.

Granted, these are not comfortable thoughts. Yet even an incurable misanthrope like Edward Abbey concedes that it was a man who composed Beethoven's Ninth Symphony. What is so strange is that we are capable, like Shiva, of both magnificent creation and eye-popping destruction. Both excellent subjects for nature writing. "We are just like squirrels, really, or, well, more like gibbons, but we happen to use tools, speak, and write," Annie Dillard writes in *For the Time Being* (2000). "We blundered into art and science. We are one of those animals, the ones whose neocortexes swelled, who just happen to write encyclopedias and fly to the moon. Can anyone believe this?"

Indeed, beyond offering the opportunity to write "natural history," that is, the story of a species, or a landscape, or an ecosystem, "nature writing" also offers one of a nonfiction writer's best chances to explore the ephemeral, the unseen, the mysterious. How does one write about the emotional life of animals, or the mysterious qualities of snow, or the ability of Inuit shamans to transform themselves into polar bears? How do we write about birth and death? Indeed, what is it, exactly, that we know about nature, about life, about our position in the world? By the time a man has lived 60 years, he has spent fully 20 years asleep. What went on all that time? We are all Rip Van Winkles. Given that the very mitochondria in our cells have DNA and RNA different from our own, writes Lewis Thomas in *The Lives of a Cell* (1974), how can we be so sure of ourselves? "A good case can be made for our nonexistence as entities," Thomas writes. "We are shared, rented, occupied. At the interior of our cells, driving them, providing the oxidative energy that sends us out for the improvement of each shining day, are the mitochondria, and in a strict sense they are not ours. They turn out to be separate little creatures. Without them, we would not move a muscle, drum a finger, think a thought." The borders we think separate us from other beings are illusory, the Buddhist monk tells us. It's all about context, relationships, interdependence. The ecologist agrees.

Indeed, in some fields, even scientists are becoming more comfortable with acknowledging the gaps in what it is possible for us to know. Chaos theory and quantum physics are just the latest examples of fields that have come to confess that the best we can do is approximate what we can know of nature's mysteries. The table on which I write this essay is made of wood, but a physicist and a Taoist master would agree that there is more empty space in the wood than there is wood. How can this be? To this and to all such questions, the Zen master Seung Sahn has an answer: "Don't know." Ask some Canadian Inuit about where their dead go after they leave Earth, and you'll get the same response. Don't know. The humility of the response is the humility with which so much good nature writing is filled. In the space between the atoms of wood lies the mystery. The best nature writing will always occupy the invigorating place between hard science and artistic abstraction. "The tolerance for mystery invigorates the imagination," Barry Lopez writes in *Arctic Dreams*, "and it is the imagination that gives shape to the universe."

34

Earth Sciences

GLENNDA CHUI

Glennda Chui writes about science for the *San Jose Mercury News,* where she covers everything from earthquakes to global climate change, nanotechnology, and the search for life in the universe. She also teaches in the science communication program at the University of California, Santa Cruz. Although she's never taken a formal course in earth science, her personal interest in the topic goes back a long way; growing up in the San Francisco Bay Area, she felt her first earthquake at the age of 3. Chui was one of a cast of hundreds at the *Mercury News* who won a staff Pulitzer Prize for coverage of the 1989 Loma Prieta quake. In 2001, she received the American Geophysical Union's David Perlman Award for Excellence in Science Journalism–News.

In August 1999, I stood in the ruins of a collapsed apartment building near Izmit, Turkey—one of 60,000 buildings destroyed in 40 seconds by the most powerful earthquake to strike a major city in nearly a century.

It was a modern building surrounded by trees and greenery. A couch and a table stood intact in a room bright with potted flowers, now open to the air. A woman's coat had been carefully draped over the remains of a wall. As the stench of death rose around us, I wondered if the coat's owner was buried in the rubble beneath my feet. I was sent to Turkey to chase the science—to bring home lessons for readers who live near a strikingly similar fault system in California. But as I surveyed the damage with a team of scientists and engineers, there was no separating the science from the politics.

Covered with a fine film of sweat mixed with dust from crumbled buildings and lime that had been scattered to prevent the spread of disease, we saw firsthand how corruption and greed had conspired with the forces of nature to kill more than 17,000 people.

Some buildings were constructed right on the North Anatolian Fault. Its mole-like tracks plowed through barracks that had collapsed on 120 military officers, a highway overpass that fell on a bus, a bridge whose failure cut off access and aid to four villages. Researchers found concrete that was crumbly with seashells, chunks of Styrofoam where reinforcing metal bars should have been.

Yet some well-reinforced buildings nicked or even pierced by the fault came through just fine, including an apartment building that moved 10 feet and had its front steps sliced off. Another home was cut in two; half collapsed, the other survived with windows intact. "How the hell?" marveled one engineer. "There's no way that building should stand in an earthquake."

That blend of science, politics, and human nature is just part of what makes earth science so compelling. It goes far beyond the academics of geology and plate tectonics to embrace earthquakes, floods, hurricanes, volcanoes, landslides—natural hazards that affect thousands of people and change the course of civilization. It encompasses oceans and the atmosphere, weather and climate change, magnetic fields, solar storms, and the way living things interact with the landscape, all coming together in one great, messy shebang to shape our world.

In fact, geophysics—the study of the physics of Earth—is not even confined to this planet. By broad definition, it applies to anything within the Sun's gravitational reach, including the planets and the primitive rocks at the edge of the solar system that occasionally hurtle in and pound species to extinction.

As a general science reporter for a newspaper, I find earth sciences stories are among my favorites—not only because the topics are so inherently cool but also because they connect with readers in a very basic way. Every part of the world has its own threat of natural disaster. Here in California, people want to know whether an earthquake is coming and when. They want insights into the geological machinery, deep beneath their feet, that is slowly grinding toward the next Big One. There's almost no scrap of earthquake science so small and obscure that the public won't wolf it down.

Elsewhere, it might be hurricanes or volcanoes or flooding, or the possibility that a favorite beach will erode away. People wonder whether the world's climate is changing, and how this might affect weather and wildlife—not to mention the value of beachfront property.

There are other matters of great importance to society: Are we running out of oil? Would a hydrogen economy have its own negative impacts on the health of the atmosphere? Are Earth's magnetic poles about to flip?

There's the slightly quirky: When photos of Osama bin Laden surfaced in Afghanistan, experts from the U.S. Geological Survey looked at the terrain in the background to try to figure out where they were shot.

And most of us love the big, impractical questions: Where did our world come from? Why is it hospitable to life, when so many planets are not? Was there ever life on Mars?

"Earth science tends to be much more concrete and easy for people to grasp. It's not as abstract as pure physics, or even astronomy or a lot of biology," says Richard Monastersky, who has covered earth sciences for *Science News* and the *Chronicle of Higher Education.* "It's not hard to sell stories on earthquakes and volcanoes and how Earth was formed and huge meteorite impacts and rovers traveling over Mars."

How to cover such an expansive and ever-expanding field?

Think Local

"Go get to know your local geologist in the local university," says David Perlman of the *San Francisco Chronicle.* "They'll tell you what journals to look at, and you'll get to know what they're doing." Establishing these relationships is critical; when an important story comes up on deadline, a researcher who's met you is more likely to call you back.

Getting to know your local disaster is equally important. Whether the potential threat to your community is hurricanes, earthquakes, flooding, or landslides, it's never too early to gather pertinent maps and reports, identify experts both near and far, and develop a plan for covering the disaster if, as sometimes happens, phone lines and power go down. Our newspaper's emergency plan assumes roads will be blocked in a major earthquake; it gives reporters and editors a strategy for contacting each other and covering the quake from wherever they happen to be. I keep hiking boots in my car, along with a backpack full of food, water, flashlight, notebook, and other emergency supplies.

As in any beat, maintain an exhaustive list of sources, organized by topic, on a computer where it can be easily searched. Get home and cell phone numbers if you can. If you're in a disaster-prone area, make a printout of essential source lists and keep them handy at home and in the office, in case your computer is unavailable. I keep a copy in my Palm and update it every few months. Find out whether local emergency officials have a system that will alert you by pager or e-mail if disaster strikes.

Take a Field Trip

Rich Monastersky has watched the youngest rock on Earth ooze like molten caramel from a volcano in Hawaii, rafted down the Colorado River for a story on how the Grand Canyon got there, and landed on the top of the Greenland ice cap in a transport plane outfitted with skis. The snow and ice extended, white and flat and empty, hundreds of miles in every direction.

"It was just incredible being up there," he says. "You get to go out to some of the most remote, amazing places on Earth, because earth scientists are not usually stuck in labs."

If you like vacationing in exotic places, he added, you can usually find an earth science story to freelance. Rich has spent parts of vacations reporting on a fossil dig in China for *Discover* and exploring the geology of South Africa's Zambezi River for *Earth* magazine.

He had promised his wife they would not raft the Zambezi. But there they were, approaching the biggest set of rapids tackled by any commercial company in the world. He grabbed the raft's safety line with one hand and his wife's life vest with the other.

"Can I possibly hold on to both Cheri and the raft at the same time? If I have to let go of one, which will it be?" he wrote. "In the end, it doesn't matter. After battering through the giant wave, we capsize unexpectedly a few seconds later. Tossed about by the water, I feel like I'm being flushed down a toilet. . . . When I finally bob to the surface, I am on my own, holding neither Cheri nor the raft."

Far from detracting from his time off, Rich says, the reporting actually makes it more fun: "It gives you a lot of insight and contacts you wouldn't have as a tourist."

But you don't have to travel thousands of miles to get out to the field. Approach scientists, find out what they're doing, and ask if you can tag along, suggests Randy Showstack, a reporter for the American Geophysical Union's weekly newspaper, *Eos*. "You'll get a better sense for who they are," he says, "and you'll have a better chance to translate their work into good, colorful language."

Quick Tips

Be careful when describing specialties. Not every earthquake scientist is a seismologist, and not every researcher who goes to sea is an oceanographer. In this age of interdisciplinary research, the boundaries between fields are often blurred; ask researchers how they prefer to be described.

Ditch the Richter Scale

This time-honored scale for measuring the energy released by earthquakes is no longer in widespread use. Other scales that are far more accurate have replaced it. The U.S. Geological Survey now uses the moment magnitude scale to describe major earthquakes. In my own reporting, I simply say an earthquake is, say, 7.1 on the magnitude scale.

Keep Up With Journals, Listservs, and Websites

For a good solid background, I recommend the following:

Journals

The publications of the American Geophysical Union, including a weekly newspaper, *Eos*, and nine scientific journals, including the *Journal of Geophysical Research*; *Tectonics;* and *Global Biogeochemical Cycles*
Geology, put out by the Geological Society of America
Bulletin of the Seismological Society of America
Bulletin of the American Meteorological Society

Listservs

CCNet: news and discussion of impacts, mass extinction, and astrobiology, plus climate change with a contrarian slant. (Sample topic: Did a comet impact trigger "nuclear winter" in the year 536, triggering a series of catastrophes that ushered in the Dark Ages?); to subscribe, contact Benny Peiser at b.j.peiser@livjm.ac.uk
NEO News: news and opinion on Near Earth Objects and their impacts; to subscribe, contact David Morrison at dmorrison@arc.nasa.gov; for more information, see http://impact.arc.nasa.gov

Web Sites

Links to earthquake information and institutes around the world: www.geophys.washington.edu/seismosurfing.html
Magnitudes, death tolls, and so on, for significant earthquakes dating back to 1556, updated as new information becomes available: http://earthquake.usgs.gov/activity/past.html (scroll down to "World: List of Significant Worldwide Earthquakes")

Meetings

By far the biggest and most comprehensive meeting is the one put on in San Francisco each December by the American Geophysical Union. It draws about 10,000 scientists who present 9,000 talks and posters. The abstract book is so thick that Richard Kerr of *Science* takes a box cutter to it, slicing it along its spine into manageable chunks. The scope of the meeting keeps expanding. One recent addition is biogeosciences, the study of how living things affect the atmosphere, oceans, weather, climate, and geology—and vice versa. Although the AGU meeting has a well-run press room, with an overview session at the start and several press conferences per day, it can be overwhelming. If you want to find stories that go beyond the obvious, you'll have to hunt hard. Some reporters do keyword searches for their favorite topics, scientists, or institutions in the online abstracts. Others phone session chairs ahead of time and ask what's new. But the time-honored strategy is to flip, page by page, through the abstract book, marking promising papers with yellow stickies. Tedious, to be sure, but rewarding; over the years I've found stories on building artificial logjams on rivers, investing in weather futures, and using the night lights of Earth to illuminate scientific questions.

Wrap Concepts Around People

One of my favorite stories was a profile of Azadeh Tabazadeh, who fled Iran as a teenager so she could pursue a fascination with chemistry—something she feared she would not be able to do as a woman living under a strict Islamic government. She told a touching story about her thirst for education and opportunity and her hopes for her own young daughters. In the profile, I tried to show not only who she was, but also how her mind works:

> [She] sits at her dinner table in Los Altos Hills, pouring a glass of Coke.
>
> Her 6-year-old daughter studies it and asks: Why are all the bubbles on the sides of the glass? Why aren't there any in the middle?
>
> Most parents would shrug. But Tabazadeh, a chemist at NASA's Ames Research Center in Mountain View, thinks about those bubbles for a long time. The result goes far beyond soda: She comes up with a startling new theory about how cloud droplets freeze—one that could profoundly affect our understanding of how people may be changing the atmosphere and damaging Earth's protective ozone.

We don't include people in our science stories nearly enough. In earth sciences, especially, we have a wide range of people to choose from: the enthusiasts who follow a space mission over the Internet, people living in the path of natural disasters. And, above all, the scientists themselves: Where are they coming from? What motivates them? What are their days like? All information that makes the science more compelling—and the researchers more human.

■ ■ ■

While covering the aftermath of the Izmit earthquake in Turkey, I tried to capture the feelings of the scientists and engineers who arrived while people were still desperately digging for survivors. "It can be difficult carrying out the dispassionate work of science while people are suffering all around," I wrote.

The scientists didn't want to interfere with rescue operations, or imply that their work was more important than saving lives or helping survivors. On the other hand, they desperately wanted to inspect cracked ground and damaged buildings before key evidence was destroyed—information that might help protect people from the next disaster.

One geologist said, "I've studied earthquakes for 25 years, and it's hard not to get excited and enthused about it. But you can't forget you're going into a disaster area. There is this aspect of it—am I doing something that will help?" He said that as he packed his bags, he found himself throwing in a pair of leather gloves—the kind you would wear for digging through rubble. "I thought, 'What am I going to need these for?' and I hope I don't," the geologist said. "But I put them in anyway."

35

Climate

USHA LEE MCFARLING

Usha Lee McFarling, a science writer with the *Los Angeles Times,* covers climate change as part of a beat focusing on earth science and planetary exploration. Before joining the *L.A. Times* in 2000, she worked as a science writer for the Knight Ridder Washington bureau, for the science and health section of the *Boston Globe,* and as a metro reporter for the *San Antonio Light* newspaper. In 1992, Usha was the recipient of a Knight Science Journalism Fellowship at MIT. She earned a bachelor's degree in biology from Brown University and a master's degree in animal behavior/biological psychology from the University of California at Berkeley.

If you plan to cover climate change, thicken your skin. The topic is one of the most highly politicized areas in science journalism today. It's not surprising, given that so much is at stake. Environmentalists fear for the very future of the planet, while conservative politicians and energy industry leaders dread pollution controls that could threaten the nation's prosperity.

As with all controversial issues, stakeholders on both sides are quick to attack reports—and reporters—that do not promote their point of view. I have been criticized by conservative think tanks for overplaying the potential dangers of climate change and scolded by environmentalists for downplaying those same dangers. It gives me solace to think that if I am aggravating both sides, then I am being fair.

Critics of climate change coverage are right to some extent. The area, in my opinion, is among the most poorly covered in science journalism. This is because politically motivated campaigns of misinformation muddy the issue and because the science of climate—both highly complex and uncertain—is difficult to convey.

Much climate change coverage exaggerates potential problems or greatly oversimplifies the issues. Reports are spotty at best, coming in droves when a particularly large piece of ice breaks off of Antarctica or there is a heat wave on the East Coast, but evaporating with the cool of autumn. Events from malaria outbreaks to species declines are attributed to climate change without adequate proof.

Climate change coverage too often falls through the cracks between beats. Climate is not only a science story. It is a political story, a foreign story, and a business story as well. It would be best if climate were covered from all of these myriad angles; more commonly, no one takes ownership of it.

Science writers, with their technical expertise, ability to translate jargon, and patience with details, are in prime position to be on the front lines of climate coverage—perhaps with occasional forays into political and economic terrain when necessary.

The topic, with its interminable feedback loops and references to past epochs, can be intimidating. But climate change—and the controversy that surrounds it—is not going anywhere for the foreseeable future. It will become even more important if and when the effects of warming become more dramatic. Here are my thoughts on the difficulties that can ensnare those who cover climate change, and how to avoid them.

The Basics

Earth's temperature is controlled by the "greenhouse effect"—the trapping of heat near the planet's surface by gases such as carbon dioxide and methane that let in radiation from the Sun but do not let it all escape back into space.

Throughout Earth's history, the climate has vacillated wildly, from the sweltering age of the dinosaurs to the Ice Ages. This variation is natural and caused in large part by changes in solar output, twitches in Earth's orbit, and fluctuations in atmospheric carbon dioxide levels.

Over the past millennium, there have been warm and cool periods that have had nothing to do with human activity. But the long-term trend of the past century is one of warming, and the rate of warming since the 1970s has been especially steep. This warming coincides with a large increase in greenhouse gases, including carbon dioxide, that have been emitted into the atmosphere from car tailpipes and industrial smokestacks.

Critics of this information abound. Some argue that Earth is not warming and that temperature analyses are wrong. Others say scientists have yet to prove that the boost in carbon dioxide has caused the warmer temperatures. Others agree that greenhouse gases have caused warming but suggest that dire consequences are not a given.

A National Academy of Sciences review in 2001 concluded that the current thinking of the scientific community is that the warming of Earth's surface in the past 50 years has likely resulted from human-produced greenhouse gases and that such warming will likely continue. The report cautioned that uncertainties remain because of the natural variability of climate and imprecision of computer models used to predict climate.

Here are some facts: The twentieth century is the warmest of the past millennium. The 10 warmest years since consistent record keeping began in the late 1800s have all occurred since 1990, according to NOAA's National Climate Data Center. As of this writing, the years 1998, 2002, and 2003 are the three hottest on record. Glaciers and sea ice across the globe are retreating. The globe has warmed 1 degree F.

Scientists project that Earth will warm 2.5 to 10.4 degrees F by 2100. The effects of this warming remain a topic of intense debate. Prospects include sea level rise, drought, and increased disease rates but also some positive aspects, like longer growing seasons. Some scientists also note the prospect of "abrupt climate change" in which the climate warms to some critical threshold and then shifts suddenly, causing radical temperature changes and shifts in ocean currents that stabilize the weather.

The Politics

For years, various groups espousing the view that Earth is not warming have hijacked the doctrine of fairness that journalists try to abide by. In other words, when a reporter quotes a scientist saying Earth has warmed and climate change appears to be a potential problem, she often will quote someone else who says the opposite. This 50/50 approach ignores the growing consensus among scientists (and even among politicians) on global warming.

I learned a similar lesson about fairness while in Washington covering a debate over whether homosexuality could be cured—an idea put forth by some religious groups. One proponent was a psychiatrist and member of the American Psychiatric Association. Almost all of the other APA members opposed his view and believed this "cure" would do harm. The association put out a position paper saying so.

It would have been irresponsible to the reader to quote the one psychiatrist who was in favor of trying to cure homosexuality and quote one who was against it. This "he said, she said" journalism fails the reader by omitting the context that the person in favor of curing homosexuality is in a slim minority among his peers. The same is true for climate change. It is important to provide the context of the larger scientific opinion. It is also key to identify the speakers on each side of the issue and, if they are speaking about science, their scientific credentials.

An ecologist concerned about species decline due to climate change is not an authority on the science of greenhouse gases. An economist concerned about regulations on the coal industry is not a scientific authority, either. You can also ask people you are interviewing to identify their funding sources as a key to their motivations. Do they receive money from the energy industry? The World Wildlife Fund? The National Science Foundation? A think tank with some political leaning?

It is important not to subtly malign those who hold minority viewpoints. Labeling someone a skeptic, a naysayer, or a fringe thinker marginalizes his or her point of view. And keep an open mind. Someone who is in the minority today may yet turn out to be right in the future. You don't want to be embarrassed in the future with articles that go overboard—like the slate of magazine articles in the '70s that warned of the coming Ice Age. Make sure to return to those with critical opinions for fresh viewpoints. Just as scientific thinking evolves, the response to it evolves as well.

Be on guard against people who want to use or deny science to push their political agenda. This is true of the oil, coal, gas, and auto industries, which fund various outreach programs fighting limits on carbon output. The same warning holds for environmental groups, which can exaggerate the impact of climate change to stoke public interest or further fundraising.

One example? Reports that polar bears could become extinct in coming decades. While the animals are stressed at the southern boundaries of their range, most polar bear experts think these predators are in no danger of immediate extinction.

Another widespread report suggested that malaria outbreaks in Africa were increasing because of climate change. A different analysis pinned the outbreaks on the diversion of public health money in Africa from malaria to AIDS. Dramatic claims require careful reporting and analysis.

How do you counter political manipulations? With the facts. Or in this case, where the facts aren't always so easily agreed upon, with the latest consensus.

The best place to get consensus thinking is from current NAS (National Academy of Sciences) or IPCC (Intergovernmental Panel on Climate Change) reports. While some will argue that these in themselves are political documents, I argue that they are not. These reports condense large amounts of current research from a wide range of sources. They are put together and approved by groups of scientists, including those critical of the mainstream scientific thinking on global warming.

Complexity

For two years, I'd been writing about ice vanishing around the world: Mount Kilimanjaro was rapidly losing its snowy shroud; Glacier National Park was set to be glacier-free in decades. And sailboats were racing unimpeded through the once ice-clogged waters of the Northwest Passage.

Now, I'd stumbled across a great story close to home. California's glaciers were vanishing too. The editors loved it. First, many of these deskbound creatures had no idea California even had glaciers. Second, with dramatic photos of ice gracing the peaks of the High Sierra, the story was a strong art package.

The report came from an ecologist at Sequoia National Park who'd led an arduous back-country expedition to rephotograph glaciers to see if they had shrunk. They had dramatically.

To flush out the story, I called a few ice experts who said they were not surprised. I found a book about California glaciers for more details. The story was basically finished, but I made one more call—a call that turned the story on its head.

California's largest glacier is in the north, on Mount Shasta. Some Web searching revealed that a scientist from UC Santa Cruz was studying the mountain's glaciers.

I reached him just as I was ready to turn the story in. He said he was indeed studying the glaciers on Mount Shasta—and they were growing. My heart sank. This new information, I thought, would make the simple, clear story I was telling more complicated and murky. What was going on?

The story—that recent warming was affecting California's glaciers—turned out to hold. His theory was that warming was leading to increased precipitation, or snow, near Mount Shasta, which was making glaciers there grow. Temperatures had not risen high enough in summer, as they had further south, to cause the glacier to shrink. I finished the story a bit more complicated, but complete—and turned it in.

The story has two lessons. First, beware of anything in climate that seems simple. With its many interrelations and feedback loops abounding, climate change is full of counterintuitive facts. The second lesson is to make that extra phone call. It would have been highly embarrassing to write a story headlined

"California's Glaciers Shrinking" without mentioning that the state's largest glacier was actually growing.

Getting to the Right Source

On August 19, 2000, I opened the Sunday *New York Times* to find a front-page story with this arresting lead: "The North Pole Is Melting." The story described two scientists who had traveled as lecturers aboard a tourist boat to the North Pole and found it unexpectedly ice free. The story described this watery expanse as something never before seen by humans, something that was not known to have occurred for 50 million years and more evidence that global warming could be real.

As I read the story, with its amazing photos of a watery North Pole, my heart sank (again). I remember thinking: "Great, the end of the world is here, on my beat, and I missed it."

I had to follow the story, of course. So I called a few sea ice experts, who are, strangely enough, numerous in Southern California. They immediately said the report was no omen of disaster. Sea ice is a relatively thin skin of ice atop an ocean jostled by violent waves. Cracks in sea ice open all the time, everywhere across the Arctic, and can be miles wide.

The fact that there was open water at the North Pole was just a coincidence—an unlucky one, in this case, for the tourists who had paid thousands to journey to and stand on the top of the world. The *New York Times* ran a correction and a revised story on sea ice, but not before suffering the wrath of David Letterman, who suggested the paper change its slogan from "All the News That's Fit to Print" to "Stuff We Heard From a Guy Who Says His Friend Heard About It."

This anecdote is not meant to denigrate John Noble Wilford, the author of the report and one of the great science writers, in my opinion. This unfortunate incident could have happened to any of us.

Two respected scientists who had been aboard the boat and who were concerned about climate change alerted the *Times* to the story. They called with what appeared to be a major scoop. But they were biologists, not experts on ice. The reporter should have called a sea ice expert—or two—before reporting something so potentially alarming. Even though it would have complicated the story to include information about the vagaries of sea ice, he still would have been first with a great story that the North Pole was not frozen—a dramatic lead if ever there was one for an accurate story about how ice is receding and thinning across the Arctic. And he could have kept himself off of *Letterman*.

Uncertainty

The people you're writing for, not to mention your editor, all want to know the bottom line: What's going to happen with climate? Are there problems ahead? If the most powerful supercomputers can't answer the question, there's no way you're going to be able to. It's not very satisfying to say you just don't know, but sometimes you have to.

You can sometimes use uncertainty to drive your narrative, looking for drama in the lengths scientists go in seeking answers to the huge questions that surround them.

In 2002, over a period of just a few weeks, two conflicting reports came out about Antarctica. One said that ice on the continent was melting and thinning; another said the ice was growing thicker. What the heck was going on? I could have ignored it until things got clearer, or written a muddled story about one side versus the other.

It turned out that the authors of the two contrasting reports worked right down the hall from each other at the Jet Propulsion Laboratory. There was a great story there about how these colleagues came to opposite conclusions, about how hard it was to get measurements in Antarctica, and a reminder that the continent was so big, it was not surprising that one end of it would behave differently from the other.

In the case of climate, more research is obviously needed to understand what is happening to Earth and how it may play out. You can report on the latest findings without having to wrap up everything in a nice package.

Getting People to Tune Back In

When it seemed to me that people were tuning out climate change as a topic, my editor and I discussed ways to get people interested in it. We went back to some journalism basics. We decided we needed to write about people who were being affected by climate, and we needed to write about the places where change was most dramatic. We also wanted to look for angles that might surprise people a bit.

For people, it was obvious: The Inuit were the most affected. So I just set off for places North—the Russian and Canadian Arctic—to see how natives there were faring. The stories were astounding: Sea ice was vanishing, whales and seals were too far away to hunt, and elders were getting lost in the tundra because the weather was so unpredictable.

As for surprises, the Northwest Passage lay open for months at a time. Polar bears were coming into towns as they waited for open waters to freeze.

Glaciers were disgorging 10,000-year-old artifacts, even frozen people, as they retreated. I couldn't believe how rich and interesting the stories were, and all there for the taking.

I'm very lucky to be at a paper that allows me to travel so widely. But it is also possible to find great climate stories much closer to home. If you're near the water, you can examine sea level rise. In the heartland, you can write about agriculture. Write about coal mining or energy production if that is a key industry in your area. Every region has water issues and animals that could be affected.

Climate change is such a hot research topic that most local universities or colleges have someone looking at one aspect of it; a profile of that work could be interesting. And much research is global, done by satellites, for example. These approaches, without geographic boundaries, are open to everyone. Just keep looking for the right person or issue to help you tell the larger story. And remember, when all else fails, people love to talk about the weather.

36

Risk Reporting

CRISTINE RUSSELL

Cristine Russell is a freelance writer who has covered science and medicine for the past three decades. She is a former national science reporter for the *Washington Post* and the *Washington Star* and appeared on television's *Washington Week in Review*. She is vice-president of the Council for the Advancement of Science Writing and past president of the National Association of Science Writers. Cris has received numerous journalism awards and is an honorary member of Sigma Xi, the scientific research society. She has a bachelor's degree in biology from Mills College. In 1987, she studied health and environmental risks as an Alicia Patterson Journalism Fellow and has been worrying about risk ever since.

Over the past three decades, the media has bombarded the public with a seemingly endless array of risks, from the familiar to the exotic: hormone replacement therapy, anthrax, mad cow disease, SARS, West Nile virus, radon, vaccine-associated autism, childhood obesity, medical errors, secondhand smoke, lead, asbestos, even HIV in the porn industry. A drumbeat of risks to worry about, big and small, with new studies often contradicting earlier ones and creating further confusion.

It's gotten so bad that some people feel like they're taking their lives in their hands just trying to order a meal at a restaurant. "Will it be the mad cow beef, the hormone chicken, or the mercury fish?" asks an imperious waiter in one of my favorite cartoons from the *Washington Post*. "Um . . . I think I'll go with the vegetarian dish," the hesitant diner responds. "Pesticide or hepatitis?" the waiter asks. The diner, growing ever more fearful, asks for water. The waiter persists: "Point source, or agricultural runoff?"

Perhaps it's time for the media to become part of the solution rather than continuing to be part of the problem. Ideally, science journalists could lead the way toward improved risk coverage that moves beyond case-by-case alarms—and easy hype—to a more consistent, balanced approach that puts the hazard du jour in broader perspective.

The challenge is to create stories with chiaroscuro, painting in more subtle shades of gray rather than extremes of black and white. Too often, as my late *Washington Post* colleague Victor Cohn once said, journalists (and their editors) gravitate toward stories at either extreme, emphasizing either "no hope" or "new hope." Unfortunately, today's "new hope" often becomes tomorrow's "no hope" (which is a good reason for avoiding words like "breakthrough" or "cure" in the first place).

Hormone replacement therapy (HRT) is a classic example of this yo-yo coverage. In the '60s and '70s, the media helped overpromote hormones as wonder drugs for women, promising everlasting youth as well as a cure for hot flashes. Concerns rose, however, with reports of possible links to cancers of the breast and uterus. Later, when the uterine cancer risk was shown to return to normal by adding an additional hormone, the publicity about HRT became mostly positive again, emphasizing its potential to protect against bone loss and heart disease. Estrogen sales soared; in 2001, sales of Premarin in the United States topped $1 billion. But in 2002, the pendulum swung back, and the headlines about HRT were all negative. Sales plummeted after a large federal study, the Women's Health Initiative, was stopped ahead of schedule because of findings of long-term risks of HRT, including heart disease, breast cancer, and stroke, that seemed to outweigh any potential benefits.

With each swing of the pendulum, the press reported the most recent study about estrogen as though it were the last word. And women were subjected to sensationalized coverage of both the risks and benefits.

We need to do better when we write about health risks. We need to write more about the self-correcting, evolutionary process of scientific research by taking readers or listeners inside the laboratory or clinical setting. At the same time, risk stories must also examine whether prudent public health policy requires action by government, industry, or individuals before the scientific answers are in.

Unfortunately, both the sources of information—public or private—and the disseminators—the media—are often unprepared to put the latest risk in context. The 24-hour news cycle puts a premium on time, the news hole puts a premium on space, and competition puts a premium on controversy and conflict over more balanced risk information. Poignant stories of individuals with claims of harm, but no scientific studies to back them up, are even more difficult. The face of one sick child often negates all the numbers in the world.

We have all been guilty, of course. I first became concerned about the problem in the late '70s, when journalists, myself included, routinely reported the latest animal test results on potential cancer-causing chemicals but often did a poor job of explaining the relationship to human health. When a 1977 Canadian study found bladder cancer in saccharin-fed rats, the Food and Drug Administration (FDA) proposed to ban the artificial sweetener. But critics scoffed that a person would need to drink 800 bottles of diet soda a day to get cancer. A *New Yorker* cartoon of two lab mice talking in a cage summed it up well: "My main fear used to be cats—now it's carcinogens." Reporters rushed from one environmental scare story to another—toxic waste at Love Canal and the nuclear accident at Three Mile Island—and relentlessly provided readers with a weekly dose of worry from the *New England Journal of Medicine* (or, as a *Cincinnati Inquirer* cartoon aptly renamed it, the *New England Journal of Panic-Inducing Gobbledygook*).

Over time, the public became more cynical, and critical, of all the press coverage. And rightly so. Many of us started to change our approach to covering complicated risk stories, asking tougher questions and seeking clearer answers about what is known about a given risk and what, if anything, can be done about it in the face of incomplete knowledge.

In writing about scientific research and numbers, it is important to understand how strong the study is, the reputations of those who conducted it, and the degree of uncertainty (for more on this, see chapter 3, Understanding and Using Statistics). Here are some additional questions that reporters should keep in mind when writing about risk:

What Kind of Risk Numbers Are Available?

Look for both relative and absolute risk information. Relative risk can often be misleading if you have no idea of what the level of risk was in the first place. Too often journal studies emphasize only relative risk. For example, *The Lancet* and the *British Medical Journal* set off a "pill scare" in 1996 when they published preliminary findings suggesting that new low-dose birth control pills doubled the risk of getting blood clots. But what exactly did "doubling" mean? A critical letter to *The Lancet* later pointed out that the absolute risk of blood clots was so small in the first place that doubling it posed little added danger—only about one additional case per 10,000 users.

How Does the New Risk Compare With Older Known Risks?

Comparing a new risk with more familiar risks can sometimes be helpful. But try to compare similar risks, if possible, that involve comparable choices or activities. Look for relevant comparisons in terms of morbidity and mortality in relationship to gender, age, geography, occupation, and so forth. If you are talking about the risk of cancer in women, for example, it may be pertinent to note that heart disease is still the number one killer of women. But be careful to provide information, not judgment. The public is often turned off when industry or government officials try to put down a new risk finding with a dismissive comment that "it is more likely you'll be hit by lightning" or "you'll get more radiation flying to California."

Is the Risk Voluntary or Involuntary?

A risk story is not just about numbers. Objectively, a smaller risk may seem less important to the experts, but not to the public—particularly if the risk is involuntary. In general, the public tends be more accepting of significant voluntary risks or natural hazards and less accepting of novel, uncertain, man-made risks imposed by others, such as radon or pollution. I remember being struck by the irony of a pregnant woman who was protesting against air pollution from a West Virginia chemical plant, while smoking a cigarette that obviously put her (and her unborn child) at far greater risk. The biggest risks are still self-imposed ones, such as smoking, eating, drinking, and driving. As Pogo once said, "We have met the enemy and he is us."

What Is the Level of Individual Versus Societal Risk?

Stories often mix the apples and oranges of risk, sounding a single alarm that fails to distinguish clearly between individual and societal risk. A risk may be relatively small to any one individual but pose a potential public health problem if large numbers of people are exposed through contamination of food, air, or water. The 1989 Alar scare, involving a probable carcinogenic chemical used in growing apples, failed to distinguish between these two types of risk. Intense publicity generated panic among parents who pulled apples from their kids' lunch boxes or poured juice down the drain even though the individual risk to any given child was extremely small. The overlooked point of the story was the government's failure to take regulatory action against a chemical contaminating a widely consumed food that posed an unacceptable level of societal or population risk over time. At the other extreme, a given health risk may be very high for a small group of individuals, such as workers exposed to asbestos, but pose little widespread risk to the population at large.

Who Is Most Likely to Be Exposed to a Given Risk?

It goes without saying that exposure, whether in the workplace, from the environment, or through personal behavior, is required for an individual to be at *any* risk, and that the greater the exposure, the higher the risk. But too often shrill messages from health officials or advocates—and the accompanying headlines—create generalized hysteria and universal concern while failing to reach those most at risk. AIDS coverage often emphasized that everyone who was sexually active was in danger of exposure to the HIV virus. While this was theoretically true, the truly helpful stories were those that stressed the risk continuum, with the likelihood of infection greatest in individuals engaging in risky, unprotected sexual acts with partners in high-risk populations in which the virus was already prevalent (such as IV drug abusers or gay men with multiple sexual partners in selected urban areas). Stressing which populations are most vulnerable to a given risk, such as children, the elderly or individuals with prior illnesses, also helps localize a risk message.

What Are the Potential Benefits and to Whom?

Too often, stories about risk drown out potential benefits, throwing out the baby with the bathwater. Journalists can't be expected to conduct risk/benefit analyses, but sorting the benefits and risks, and the degree of certainty, is useful. In 2004, news stories about claims that antidepressant drugs may increase the risk of suicide in adolescents sometimes failed to present the bigger benefit picture, namely, that increased use of antidepressants is strongly linked to a significant drop in overall suicide deaths among young people over time. More subtle coverage explained the ambiguities of balancing individual risks and benefits against the societal consequences. In some cases, potential beneficiaries and at-risk groups are different. Workers and residents living around a chemical plant may face higher exposure risks, while the consumers who buy the company's products receive the benefits.

What Can Be Done About a Given Risk, If Anything, and by Whom?

Stories should outline what government or industry can do to prevent or mitigate risk, as well as what individuals can do to protect themselves. Time is the crucial factor. Some risks are historic, with the damage already done. Other risks require complicated preventive or corrective actions by the public or private sectors that take time as well as political or legal prodding. In the meantime, individuals face difficult choices about how best to manage their own risks when the scientific answers are not yet in. Prevention is an overused word and is not always applicable. The largest known risk factor for breast cancer, for example, is a strong family history, which is not controllable. Until now, the only action was increased medical follow-up. However, rapid advances in genetic research are creating new options that may change the risk equation for cancer and other diseases and raise new ethical dilemmas that are the fodder for further stories.

■ ■ ■

The credibility of the media, and ultimately of the scientific enterprise itself, is at stake in our coverage of risks to human health and the environment. There are many "publics" out there, with widely varying intellectual and emotional backgrounds. Some people become hysterical about virtually everything, while others give up or tune out because of information overload. In between are the readers and listeners looking to the media for some guidance in understanding the risks we face and how to deal with them. Sometimes the best we can offer is the simple truth that science currently has no clear answers, so we need to learn to live with uncertainty. We owe it to our audiences to provide more sophisticated, balanced risk reporting that goes beyond the "fear factor" approach, to put Chicken Little back in her place.

Taking a Different Path

Journalists and Public Information Officers

■ ■ ■ ■ ■

SIMILARITIES AND DIFFERENCES

All science writers share some common goals. Whether they write for a news organization, university, medical center, museum, nonprofit, government, or industry, they find out what's new and intriguing, or new and useful, or new and just plain fun to write about, and they explain it in language that nonscientists will understand. The National Association of Science Writers feels that this wide array of science writers shares enough in common to belong to the same national professional organization.

And so it is that you will find among our NASW members reporters who work for newspapers, magazines, television, radio, and the Internet. And you will also find members who are science writers, public information officers (PIOs) who do some science writing, PIOs who do no science writing, and public relations practitioners who work for universities, medical centers, organizations, government, and industry. There are also freelance journalists who contribute articles to the media and write books, and freelances who write on a work-for-hire basis for universities, medical centers, organizations, government, and industry. Some freelances do some of each.

When they cross paths, journalists and PIOs usually work well together. A PIO can be a blessing to a reporter who is looking for an expert but is not quite sure who at an institution or within a professional organization or government agency is knowledgeable about the science and about how to talk to reporters. Journalists also respect a well-written press release on a newsworthy topic that doesn't yell "sell!" from the first sentence.

But as surely as a baseball team manager will occasionally and predictably get in the face of an umpire over a call in a big game, there will be times when a reporter and a PIO will be at odds with one another in small and big ways. And, just as on the ball field, it can get nasty. The clash comes when a reporter, trying to find out the truth about a matter, seeks information or sources and runs headlong into a PIO with conflicting motivations. Both want to give the public the truth. But the PIO also has an additional function, to look out for the best interests of an employer, which may mean that the PIO may try to block access to sources, withhold information, or try to manage it.

In the next section of the *Field Guide*, we feature chapters from some of the best PIOs in the business, who describe the opportunities and exciting challenges their jobs offer. A couple of these authors also describe practices that go with those jobs but that could cause journalists to cringe. Joann Rodgers acknowledges in chapter 38 that there are times when journalists may feel that PIOs are stonewalling, but PIOs don't see it that way. The PIOs, Joann says, are trying to get the facts out as soon as they believe they have all the facts they need. "PIOs worth their salaries will do all they can to push for speed and full disclosure," she writes. Yet in the same sentence, she warns, "but as someone who has been in the trenches for more than 20 years, I can promise that the gap between what journalists want and what institutions will provide will never completely close."

Colleen Henrichsen brings up another rule in chapter 39 that a number of PIOs have in place but that many journalists find, especially when on deadline, is more an obstruction than a help. "Agencies differ on procedures for handling media inquiries," Colleen writes. "Some require scientists and administrators to refer reporters to the institutions' communications office. NIH encourages scientists to contact their institute's communications office."

Some PIOs also sit in on an interview between a journalist and one of the institution's scientists, or even listen in on a phone interview. This is another practice that troubles us. Reporters who are asked to allow this practice tend to worry that sources will be less candid if a representative of the institution is listening to their conversations with reporters. Many reporters will refuse to abide by such a rule.

To be fair, we should mention that reporters are not altogether altruistic in their aggressive requests for information. Journalism is very competitive, and today's 24-hour cable news, Internet news, and blogs all raise the level of pressure that the rest of the media feel. Competition to be first, to be best, and to win prizes has led to some bad practices by a few reporters and editors, practices that we condemn.

Yet competition and timeliness are intrinsic to news operations. And when they're handled professionally, the public benefits. So we champion the right of the press to get information as quickly and fully as possible. We reject any guidelines that have the effect of forbidding direct and unfettered press access to all information, as long as the information sought does not violate a patient's right to privacy. Reporters must continue to push for access to all sources that can supply information for stories the public has the right to know about. In disputes over access between journalist members of NASW and PIO members of NASW, we editors, who are all journalists, come down not surprisingly on the side of journalists.

Nevertheless, we have given PIOs their voice in part VI for two reasons. One, we think it is important for students who may be interested in pursuing a PIO career to get a sense of what would be involved. We want to make the *Field Guide* as comprehensive as possible, a true reflection of the myriad ways there are to be a science writer in the twenty-first century. And two, we think it's important for journalists to clearly understand how PIOs do their jobs.

And so we put part VI forward with the recognition that some of the practices recommended here will not always meet with approval and acceptance from journalists. We caution that the mission of PIOs sometimes runs counter to the mission of journalists, and we urge all young journalists reading this book to become champions of one of the principles on which NASW was founded: the promotion of full and free access to the news.

THE EDITORS

Part VI

Communicating Science From Institutions

■ ■ ■ ■ ■

"You'd better come on over to the lab. We've got a problem."

Those are words no public affairs officer ever wants to hear, especially two days after announcing to the world that researchers at his institution had confirmed the controversial claims of cold fusion. At the lab, researchers who had confidently told the TV cameras about detecting excess neutrons in their experiment—the hallmark of nuclear fusion—were sheepishly telling me that those supposed neutrons were actually only errors in their instruments.

Standing around in the lab, we quickly agreed we had to tell the world about our mistake with the same level of effort with which we'd earlier announced our confirmation. We expected that would likely mean, in the popular sports euphemism, career-ending injuries for all of us.

But it didn't turn out that way. For reporters covering what in hindsight can only be described as an aberration of science journalism, our retraction provided a new lead that gave the story extended life. To my surprise, we did not become outcasts. Fifteen years later, I still work with some of the editors and writers I was sure would never take phone calls from me again.

But I've never again called a rushed news conference to announce surprising research findings, and I've also never again had to write a press release retracting a claim.

That experience taught me that a major part of my job is to be a naysayer, a doubting Thomas, a wet blanket, a jug of cold water, and a fire extinguisher. From the cold fusion fiasco, I realized that I could have served my institution better by challenging those who wanted to rush into an announcement. I should have argued for some real peer review and more time to consider whether we were really seeing neutrons—or error from instruments designed to measure the effects of atomic bombs. And I should have warned more forcefully about the consequences.

In chapter 37, Earle Holland argues that university public information officers work for the readers who invest time reading their copy and for the people they write about. I'd add a third boss: the long-term good of the institution. Like children, sometimes institutions want to do things that we know will likely hurt them. Like a parent, sometimes we have to take them aside and have a serious heart-to-heart talk.

But to do that, we must have trust and respect from those who make these decisions. For those of us with backgrounds in communications, earning trust and respect from faculty and administrators can sometimes seem more difficult than earning trust and respect from writers, editors, and broadcasters.

For instance, not all administrators agreed with the decision to immediately retract our cold fusion claim. A heated discussion was taking place among two administrators even as television crews gathered to hear our confession. The final decision lay with a vice president who knew what was at stake. Had we not made the announcement, the news would have dribbled out anyway. We would have lost our trust, the institution's reputation would have been harmed—and I wouldn't be writing this introduction.

In chapter 38 on institutional relations, Joann Rodgers describes communicating in a crisis. Those of us working in nonmedical institutions face fewer life-and-death crises of the kind she describes, but the ones we face can really be biggies, like cold fusion. How well we deal with them determines whether our institutions retain trust—and, ultimately, whether we are successful as professionals.

Also in part VI, Colleen Henrichsen (chapter 39) describes the unique challenges—political, organizational, and otherwise—of science writing within government agencies, and Frank Blanchard (chapter 40) describes the issues of writing for philanthropic and other nonprofit organizations in these times of heightened budgetary competition. Mary Miller (chapter 41) discusses the challenges of educating and entertaining museum visitors while writing very sparingly. Finally, Marion Glick (chapter 42) describes working in the world of corporate communications, an area where business considerations make it even more complex than university, government, museum, and institutional communications.

In a 30-year career that began before the Web, e-mail, and digital cameras, I've learned more than skepticism for dramatic claims. Following is my top 10 list of what's most important for public affairs science writers to know, whether they work for universities, the corporate world, government agencies, museums, or institutions.

1. *Take a long-term view.* The people you work with in your institution and in the news media could be your colleagues for a very long

time. Treat them in such a way that you'll still be able to work with them next week, next month, and next year. When Dean Getthewordout insists you call the *New York Times* with details of his departmental reorganization, think about what that could mean for your relationship with the *Times* when one of your researchers publishes a truly newsworthy finding. And remember that the writer from the small publication whose phone call you don't return today may be a writer at *Newsweek* tomorrow.

2. *Think visually.* Broadcast outlets need lots of moving images to cover even a brief story. Doing the legwork ahead of time—lining up patients to be interviewed, getting signed approvals, and knowing what you can and cannot show the camera—is essential to working with harried TV producers. Websites increasingly need animations and diagrams as well as videos and stills to describe complex topics. And even the traditional print media needs imagery to explain the story to readers who are increasingly visual. Finding the visual component for your story is as important as finding the words.
3. *Understand those with whom you work.* If you've never worked in a daily (or hourly) newsroom, struggled to help fill a 60-minute daily newscast, put together a monthly magazine, or juggled multiple story deadlines as a freelance, get to know people who have. You'll be better able to meet the needs of daily newspaper and wire service writers, broadcasters, magazine editors, and freelances if you understand what drives them.
4. *Know your institution.* This seems obvious, but how to do it is not. For small institutions, it's possible to know all your faculty and what they're doing. At larger institutions, get to know those people who know what your faculty members are doing. They may be development officers, assistant professors building new research areas, or associate deans who understand what the communications office can do for them.
5. *Keep the trust.* Rick Borchelt, director of communications at Berman Bioethics Institute, Johns Hopkins University, argues that one of an institution's most valuable assets is the degree to which it has public trust—what he calls a "trust portfolio." Science writers help maintain that portfolio, which is fragile and difficult to repair if broken. Take good care of your institution's trust portfolio.
6. *Maintain the highest ethical standards.* After what the communications world has been through in recent years, the message is clear that outright lying and misrepresentation are not only

wrong—they're also counterproductive. But being ethical also includes how we treat people. If you want to be treated fairly, treat others that way. This applies to the special deals, exclusives, and embargo "leaks" that may be tempting.

7. *Know what news is.* This is easy for those who've worked in the news media, difficult for those who haven't. Some unfortunate PIOs may believe that news is what their institutions deem it to be. They face an unhappy future. To learn what news is, study what the media covers in print, on the Web, and in broadcasts. In science writing, the gold standard for newsworthiness relates to publication in journals such as *Science, Nature,* or the *New England Journal of Medicine.* Presentations at prestigious conferences also count, as do receiving patents or achieving some other "news peg." A professor's grant running out does not constitute news.
8. *Work in the overlap.* Picture a diagram with two intersecting circles. The circle on the left represents the information your institution would like to get out to its constituents. The circle on the right represents what news outlets consider newsworthy. Your job is to understand what's in the area where the circles overlap. That's where you should spend most of your time. At times, however, you'll have to work in the right part of the circle dealing with issues the media wants to cover that your institution doesn't want its constituents to know. How well you handle that may well determine your success in other areas.
9. *Earn respect.* If you work at a university or other institution, chances are you don't hold a Ph.D. and haven't published articles in *Science.* How do you earn the respect of faculty who do and have? By being honest with them, doing what you say you are going to do, involving them as partners—and respecting their time and knowledge. Find an ally in the administration who understands communications—or who at least will listen to you. Build relationships of trust early, before you face a crisis or must fight a communications battle.
10. *Be a lifelong learner.* We all come to science writing with different backgrounds, but none of us with an understanding of what will be discovered in the decades ahead. Keep your knowledge current by joining professional organizations, reading the top general magazines, science publications, and trades, participating in professional development opportunities—and talking with really smart people. Even if your background is in the news side of journalism, consider courses and workshops to hone your news writing skills.

The chapters that follow contain advice and observations from some of the best practitioners in diverse areas of public information science writing. Even if this is not your career choice, their words may help you understand those areas and provide fresh insights for your own.

JOHN D. TOON

John D. Toon has been manager of the Georgia Institute of Technology's Research News & Publications Office since 1997. Prior to that, he was a senior science writer at Georgia Tech and manager of news and information for the university's Advanced Technology Development Center.

37

Universities

EARLE HOLLAND

Earle Holland has headed research communications at Ohio State University for more than a quarter century. He's served multiple terms on the board of the National Association of Science Writers as well as on the board of the Society of Environmental Journalists and on the national advisory committee for EurekAlert! A former reporter for the *Birmingham (Alabama) News,* he wrote a weekly science and medicine column for the *Columbus Dispatch* for 18 years; for the last six years he has written the national weekly column *GeoWeek* distributed by the New York Times Syndicate. Earle also taught a graduate science reporting course at OSU's School of Journalism for 20 years; the OSU research communications office has won more than 65 national awards while under his direction.

Science writing at a university has to be one of the world's great jobs. If the institution is serious about its research, you're a kid in a candy store. In my case, at Ohio State University, with more than 3,500 faculty, the question is what to write about first—not where to look for stories. Big universities are that way, but the same rules apply for smaller places that are intent on doing great research. Let's begin with the basics.

What Is the Job?

While public information officers at universities face a buffet of varying tasks—from covering boards of trustees' meetings to athletic scandals to student riots

—the role of the science PIO is more focused: Concentrate on university research; explain what is new and why it is important to the public. Stated that way, the job seems simple, but science writers at a university may have to jump from astronomy to immunology to psychology to anthropology all in the same week. That represents a lot of intellectual gear shifting; but remember, the rules about reporting on research generally stay the same from field to field. What is the news? Why is it important? What is the context for the research? That is, what are the questions that drive it? Why should the readers care? And last, do the findings point us somewhere new? The only things that change from story to story are the researchers' language and the culture specific to their fields.

Where Are the Stories?

Nearly every time I give a talk on university science writing—and there have been dozens—someone asks the classic question: How do you find your stories? The glib answer is "Everywhere;" but in truth, that's pretty accurate. Some people envision situations where top researchers have a "eureka" moment and then immediately get on the phone to the campus science writer to get the word out. Or perhaps the researcher's department chair or dean, ever attuned to their colleagues' work, is the one to pass along such news. I wish that were so; but sadly, it's more likely that researcher, department chair, or dean will never think about calling a writer until long after everything else is done.

Others think that science writers learn about research advances osmotically—information simply wafts its way across campus until it is picked up by the writer's antenna. In a sense, that's perhaps closer to how it actually works. At Ohio State, we've adopted pretty hard and fast rules about stories, what they contain, when we focus on them, how we find them. More than 95 percent of our stories are linked to reports to be published in peer-reviewed scientific journals, or are presentations scheduled for major national or international scientific meetings. Tying our reportage to these forces us into a symbiotic relationship with the embargo system and puts us in the same mind frame as the science journalists who ultimately receive our stories. Not every story is embargoed, but the news hook for the story is usually that publication or presentation.

Long ago we decided to avoid science "features"—overview stories that cover a researcher's work without focusing on discoveries or advances. While we love doing these stories and agree that they can help explain the science, they lack the immediacy that "news" requires, and the likelihood that they'll stimulate media coverage is pretty low. Likewise, we avoid the classic "grant announcement" story, where Professor X has received a grant for umpteen thousands of dollars and plans to study a scientific problem. Doing those, we feel, places the emphasis on the researcher's getting money rather than on what the results can bring to society.

We end up canvassing a host of sources looking for stories. We scour the top-tier general science and medical journals, of course, but also the key journals in dozens of specific disciplines. We do daily searches of journal databases looking for academic papers written by our faculty. We comb through daily news coverage looking for stories that we might have missed. And, of course, we constantly interact with our researchers. But to call ours a true beat system is pretty far-fetched—the university has more than 100 departments, and there are only four science writers to cover them all. One of us covers clinical medicine and life sciences, another physical sciences and technology, a third watches the social sciences and humanities, and I dabble in an eclectic mix of various fields. We all focus on areas we understand, although we sometimes might overlap.

The key to success is in the relationships we establish with individual researchers, and then research teams and ultimately departments, based on our past reporting of their work. If we get it right in their eyes, they trust us and are more likely to work with us for years to come.

Who Is the Audience?

I've always believed that a good university science writer is not much different from a good newspaper science reporter. The first allegiance must be to the readers—what will interest them, what do they need to know, and what do they have a right to learn? That approach has endeared us to the reporters who receive our stories. They expect us to work as they do, and in turn, they use our stories as starting points for their own—which is literally what our mission happens to be.

The problem with this approach is obvious: If the institution's leadership doesn't agree with the logic, then the university PIO's allegiance to the university may be questioned. In my case, this approach has worked well for decades so it isn't often questioned. People with less time on the job may need to passionately argue the case for this approach.

The university science PIO's second allegiance must be to the researchers themselves, and to the work that they do. Good science communication on a campus is the product of a partnership between the researcher and the writer. Both have to contribute equally, and both must have shared goals. That's not always easy when the partners represent different cultures. While both believe in accuracy and "truth," disagreements can arise over what approaches should be taken to tell the story and explain the research. The writer obviously must defer to the scientist for technical accuracy, but the converse holds when it comes to storytelling. It is a tough dance to pull off well, but once accomplished, it charts the course for future stories.

But things don't always go smoothly . . .

When I was at Auburn University and still new to writing for universities, the National Science Foundation asked us to produce a short booklet touting one of our projects. Two industrial engineers were looking for the best way to convey technological advances to state legislators. A couple of interviews and a mountain of reading later, I sent the researchers a 15-page draft to review. What followed was a back-and-forth comedy with each of their revisions growing longer until the text reached 60 pages.

"This is as concise as we can make it," they argued. Fortunately for me, they'd earlier done a journal article that was short enough. A quick edit for style and dumping of the jargon, and it was done. NSF was thrilled with the finished product.

My faculty, however, were livid. They wrote the university president, reviling my actions and accusing me of "descholarizing" their research. The president, in turn, commended my work and gave me a raise. The faculty eventually got over it, and I learned a good lesson about working with researchers: Sometimes the best way to do your job is not to do what you're told. Instead, do what's right.

That lesson is as valid now as it was three decades ago.

The Emerging Public

We all talk a lot about writing for the public, but in truth, PIOs have historically focused on intermediaries: reporters, who are gatekeepers to the public, or researchers, who are gatekeepers of the information. Only now, with the advent of the Internet, can PIOs seriously consider themselves as reaching out directly to the public. Stories we produce about our research advances are literally at the fingertips of the entire world, or at least that portion of it with access to computers. Of course, institutions that produced research magazines have reached readers directly, but usually the numbers of potential viewers for those stayed in the tens of thousands or less. Now we can reach millions!

It is important to remember that most of those readers can readily tell the difference between good journalism and hype. So the obligation is on the science PIO to function even more as a science reporter, rather than as a public relations practitioner. The common belief has long been that it takes 10 years to build a reputation and only one year to lose it. The truth, however, is that reputations can be lost much more quickly. Research universities must be seen as places to be trusted, where researchers are working for the common good and integrity is the coin of the realm. The easiest way to achieve that is a commitment to telling the truth, and that is best portrayed through good science journalism.

Public institutions, especially, are answerable to the populace, and if the public sees the university as an institution that fosters new knowledge and inquiry for the greater good, then it is much more likely to support the university's role in society. And the public's perception can be affected by how we report on research.

In the early 1980s, AIDS wasn't yet a recognized disease, but researchers were puzzled by patients who oddly faced both an opportunistic pneumonia and Kaposi's sarcoma. The student newspaper at Ohio State one morning reported that one of our researchers claimed to have found the cause—a microbe native to the Caribbean and Africa that he said he had cultured from a batch of factor VIII, the blood-clotting element often missing in hemophiliacs. He lamented that he couldn't solve the mystery because his grant funding was running out. Within an hour, the local Scripps Howard paper called, and then the Associated Press, and finally a CNN reporter saying he and a crew were boarding a plane and expected to be on campus in about three hours.

While my boss was elated with the coverage, I explained that the work hadn't been published in a journal or presented at a scientific meeting. The researcher had refused to provide his data to officials at the Food and Drug Administration. I even learned that the researcher had called federal officials trying to leverage more grant funding based on the press reports.

"We don't report on research this way," I explained. "Nobody does." I argued for issuing a release disowning the reported findings, pointing out that they had not undergone any sort of peer review—the benchmark against which research must be judged. I added a couple of statements about our surprise at the announcement and we released our statement by late morning. The resulting coverage was fair. About a year later, the researcher left campus for parts unknown.

Our decision gained the university years of credibility by taking the high road, publicly announcing the standards our research must meet, and in doing so taught readers a bit more about the scientific process. My job was to defend the integrity of the institution and of the research community, not to capitalize on a media coverage opportunity.

The Challenge

More research emerges from university campuses than from any other source in society. For science PIOs, that presents an opportunity that is both wonderful and monstrous. It is wonderful in that we can literally be at the threshold of the newest science. We can look over the shoulders of world-class scientists, share their excitement, and be the first to report on major advances, while in the process becoming long-term friends. The monstrous part comes with our responsibility to get the story right. Doing that requires balancing the needs of journalism with the constraints of science. It also means withstanding the very real temptation to embellish, and the often-fierce pressures to exaggerate, that may come from those who run the show.

Ultimately, we measure our success by how well we tell the science stories and by trust—the trust of researchers in our accuracy, the trust of our institutions in our work, the trust of the gatekeepers who feed off our offerings, and the trust of the readers in what we say. Like peeling layers off an onion, the stories at a university are seemingly endless.

Good hunting.

38

Institutional Communications During Crisis

JOANN ELLISON RODGERS

Joann Ellison Rodgers, a former president of the Council for the Advancement of Science Writing and the National Association of Science Writers, is the author of six books of nonfiction, including *Psychosurgery: Damaging the Brain to Save the Mind* (1992) and *Sex: A Natural History* (2002). Winner of a Lasker Award for Medical Journalism, she is a fellow of the American Association for the Advancement of Science and one of only 24 nonscientist members of Sigma Xi, the scientific research society. After 18 years as a reporter, columnist, and national science correspondent for the Hearst Newspapers, she became director of media relations for Johns Hopkins Medicine, where she also serves as deputy director of public affairs.

Shortly after I left daily newspapering in 1984 for a post in Johns Hopkins Medicine's public affairs office, I was called to a meeting of senior administrators at the Johns Hopkins Hospital. The assignment was to decide what to say publicly—or whether to say anything at all—about an outbreak of deadly meningitis in the newborn nursery, and the need to close it until state and hospital epidemiologists had tracked down and eradicated the source of the infection. The right things were already being done to protect the public and the workforce, to take responsibility for the problem, and to investigate and fix what might have gone wrong. The issue was communications.

My still-fresh reporter's instincts led me to propose that Hopkins call a press conference to tell the bad news quickly, before it leaked and the press suspected a coverup. We would publicly advise prospective mothers-to-be that Hopkins would arrange for their deliveries at other institutions.

Despite worries that press coverage would hurt our reputation, scare patients and visitors, and invite lawsuits, I got the benefit of the doubt and personally broke the news on camera that same day. Hopkins was rewarded with a newspaper editorial praising us for putting patient safety first, a bolstered reputation for credibility, and a sure bet for increased referrals and revenue.

Not a bad outcome, although not a great one, either. I might have asked that a physician or nurse deliver the news, putting a bona fide expert's face on the story. (The press corps wasn't exactly thrilled with my "credentials.") I could have made sure insiders got a "heads-up" advance notice before they saw my face on the 6 p.m. news. (They grumbled—appropriately—about having been blind-sided and ill-equipped to answer follow-up questions from patients, families, and journalists.) And I should have alerted public information officers (PIOs) in the state health department that they would surely get calls from the press as well and should be prepared to respond quickly.

Still, 20 years later, the option of whether to communicate or not communicate during a crisis remains widely recognized as no option at all. (Think Exxon *Valdez* and Three Mile Island.) And, in this regard, my PIO's instincts are the same as my reporter's instincts: Tell bad news first, fast, and fully. This is the PIO mantra and should never change.

What has changed is the complexity of issues surrounding medicine and the health care system, along with more instant demands for scrutiny and accountability, howls over safety lapses and malpractice, mounting regulations, and the "corporatization" of health care delivery. Crisis management at academic medical centers is now a full-time enterprise for platoons of professionals. Institutional debates and decisions over how, what, and when to communicate about crises require systematic and sophisticated planning, management, and techniques.

It follows that media and public relations experts need particular skills and resources to handle communications when an institution's reputation, revenue, and core missions are threatened by events.

What Are the Crises?

To begin with, we need a big-tent view of what constitutes a crisis. Some candidates are easy to recognize by the headlines they have made:

- A young patient at Duke University gets a mismatched organ transplant and dies.
- A healthy research volunteer at Hopkins, who also is an employee, has an unexpected reaction to a chemical and dies, leading federal regulators to shut down thousands of clinical trials affecting thousands of patients.
- A whistleblower at Hopkins claims medical residents' work hours are in violation of regulations, prompting an accrediting organization to consider decertifying an entire residency program, threatening the careers of hundreds of residents, and panicking medical students waiting to "match."
- A respected research team at Hopkins withdraws a scientifically and politically controversial paper, already published in a prestigious journal, after the team itself discovers an error in its experiments.
- Several medical centers agree to repay Medicare millions to settle what the government insists were fraudulent claims.

Other crisis epidemics are emerging over faculty conflicts of interests and "effort reporting" violations. These days, it's difficult to find a scientist *without* an interest in, or relationship with, a commercial enterprise, inviting widespread assumptions of greedy wrongdoing. Highly productive investigators with multiple federal grants who historically were trusted to approximate the amount of time spent on each endeavor must now bill and account for every hour and face fraud charges if they can't produce precise documentation.

Sometimes a crisis that suddenly occurs grows out of work that has been ongoing. The exemplar here is the seemingly "sudden" epidemic of medical errors, underscored by the Institute of Medicine's two blockbuster reports in the late 1990s on the high rate of serious errors in hospitals. Hospitals acknowledge that errors occur and also recognize that multiple drugs (the average inpatient gets more than a dozen), shortened lengths of stay, lack of insurance, and a sicker inpatient case mix are increasing the risk; and most are committing huge resources to address this serious problem. But the press and public mainly lack context for what's going right as well as what's still going wrong.

Advances in biomedical research and clinical medicine will create long-running, if episodic, communications crises. Think stem cell research, assisted reproductive technologies, animal experimentation, genetic fingerprinting.

On September 11, 2001, a special category of crisis emerged, requiring sensitive handling by communicators. The communications issues to be dealt with during these crises sometimes can leave even old pros reeling. Should reporters' questions about the existence and whereabouts of labs that work with anthrax or other biological agents be answered if terrorists can read all about it, too?

Does the press have the right to know if any faculty or students are being questioned by the FBI about terrorist events?

Anticipate Needs, Demands, and Criticism From the Press

Complicating every category of crisis are new technologies and practices in journalism. Ten years ago, it was miniaturized hidden cameras. In vogue today are the Internet and Web journalism, which have demolished the concept of conventional deadlines forever and sharpened the appetite for instantaneous response to requests for information, sometimes literally within moments of an event. This Internet and Web journalism, along with Web-savvy, Freedom of Information Act—ready investigative teams, have, for better and worse, put intense pressure on institutions and their spokespersons to produce faster, more comprehensive reports, statements, interviews, experts, and backgrounders. Within the limits of our budgets, staff, and access to information, we need to provide them.

It's worth noting that a communications crisis can emerge even when the subject itself is benign, or over what at first blush seems good news. And in the land of communications, no good deed may go unpunished. The ABC News prime-time series *Hopkins: 24/7*, made by 18 reporters and producers who spent three months, around the clock, filming Hopkins medicine in action, warts and all, was seen by millions and won an Emmy. But even it became a symbol of betrayal and target of anger. The Joint Commission on Accreditation of Health Care Organizations publicly (and incorrectly) doubted that patients filmed by ABC had given written informed consent, and wrote new rules that have made it all but impossible for Hopkins and other institutions to allow such access to any other news organization. Rival TV networks and newspapers demanded to know how much we paid ABC for the coverage (none at all, and ABC paid for every last meal and phone call for its staff) and blasted Hopkins for giving ABC—and not them—unprecedented access.

What all of these events have in common is this: the potential to one degree or another not only to draw short-term negative reaction, but also, over the long term, to have unintended consequences that demoralize the workforce, depress recruitment of faculty and top students, stifle philanthropy, damage reputations, and undermine core missions.

However much the press thinks that the immediate, harsh, intense spotlight it can focus on an institution is *the* critical driver of institutional crisis communications, the reality is far more layered and long term.

PIOs worth their salaries will do all they can to push for speed and full disclosure, but as someone who has been in the trenches for more than 20 years, I can promise that the gap between what journalists want and what institutions will provide will never completely close. I'm reminded of the criticism Hopkins sustained from some reporters when our institution declined for a few days to go public with the name and address of the young, healthy research volunteer and employee, Ellen Roche, who, as noted earlier, died while participating in a baseline physiological study using a challenge dose of a chemical called hexamethonium to simulate an asthma attack.

Hopkins duly reported the incident when it first happened, to regulators and other agencies, and later, but relatively quickly, to the press as well in lengthy statements, interviews, fact sheets, and updates. But news organizations were furious, complaining we were not fast enough about releasing certain details. Even though we promised to make public results of internal and external investigations of the event—and we did—the frequent complaint was that we were "stonewalling" in order to avoid lawsuits.

In fact, lawsuits or settlements are pretty much assumed in such situations. The family and its representatives asked that their privacy be honored and that we say nothing to the media. Over the next weeks and months, the family did not speak to the media either, further fueling reporters' convictions that Hopkins was "hushing them up." Hopkins needed to address both the family's grief and the problems with our research oversight processes that the tragedy uncovered. That took hundreds of hours from dozens of individuals, whose first job was to get the thousands of research protocols, shut down by the Office for Human Research Protections, back up and running. Clinical trials involving thousands of people in need of treatment were at stake.

The media's frustration was understandable, but it is an indisputable fact of life during institutional crises that some things will be withheld from the press—some for a while and some forever—for legal as well as humanitarian reasons. Other information will be withheld only until there is some degree of certainty about what the facts are to be reported.

Institutional communicators (like institutional leaders) who fail to understand the complex links between crises and their *long-term* consequences don't last long in their jobs; reporters who fail to understand the institutional dynamics at play at such times may tilt at windmills, making an appreciation of the elements of modern crises a worthy pursuit for insiders and outsiders alike.

TWELVE RULES FOR COMMUNICATING IN A CRISIS

What should happen when an event triggers institutional response machinery? What skills, strategies, and tactics are needed and work best to get the story, and get the story out? (These are not the same thing at all.)

Here are some best practices for gathering, organizing, and distributing information during a crisis, based on field-tested plans and experiences shared graciously by many practitioners of PR and media relations:

1. Have a seat on your institution's crisis management team. Being summoned at the last moment to issue statements blinds you to the nuances and questions that may come back to haunt you. As a surrogate for what the press and public will want to know, you can help focus the crisis management team on the big picture and the details when it comes time to develop messages and statements.
2. Develop tailored, written strategic and tactical plans for use by the crisis management team, all based on facts, and on both short- and long-term goals and consequences. Is this a one-day sprint of a story or a likely marathon? Was the institution at fault, or do the facts show otherwise? Is there a public health component? Should there be a written statement issued only? (The more complex the issues, the better this option, because written statements can be carefully nuanced and don't "mutate" as oral communications often do.)

   Will there be a press briefing and interviews? If so, who are the best spokespersons and how much training do they need? What are the key messages and key audiences, internal and external, for each? Is there value in recruiting "third party" supporters and spokespersons? Should there be letters to the editor and op-eds?
3. Make clear to reporters any external forces that may circumvent full or early disclosure of information. There often are legal, regulatory, policy, moral, or ethical reasons why certain information cannot be disclosed, ranging from Health Insurance Portability and Accountability Act (HIPAA) regulations, privacy rules, and legal settlement terms to requests from patients and families and public safety considerations.
4. Question internal sources of information about everything. Be skeptical. Reporters will, and you need to be comfortable with the credibility of the information you're being asked to divulge and share. Be not afraid to ask!
5. Offer to draft all statements and circulate for additions, corrections, and interpretations. This keeps you in direct touch with the appropriate institution representatives so that no one else at the institution who may want to prepare a statement gets between you and those you are quoting. It's a lot of work, involving sometimes dozens of drafts and long hours. But it's worth every minute because you need to be able to stand behind the statement.
6. Anticipate every conceivable nasty, hostile question your institution's experts might get, and then persevere until you get answers. Even if the best you can do fast at first is "we don't know but we'll find out." Fashion and keep a rolling list of questions and answers during a crisis. These and the statements will serve as the foundations for internal and external communications, as well as subsequent stories for internal publications, dean's letters, and letters to patients, students, families, and donors.
7. Find and fix holes (preferably before a crisis) in your communications staffing and hardware and software. A crisis is no time to run out of pager batteries, figure out how to access your Virtual Private Network from home, or decide where to put a fully equipped auxiliary press center.
8. Assume that anything you say internally is going to find its way outside, and behave accordingly. Label drafts and e-mails "Privileged and Confidential" and then be very careful what you say on e-mail.
9. Have a round-the-clock, on-call service for communications. No exceptions. Create special online and print templates for crisis communications. And keep finalized crisis statements in a file that all staff can access from any computer. This can greatly expedite off-hours information for press. Uploads of new information to websites and intranet sites should be prompt.
10. Network. Get to know—really well—your crisis communications team members. Know cell phone and pager numbers, home telephones, and vacation schedules of top corporate officers and deans, as well as general counsel, campus security chief, dean for policy coordination, compliance officers, hospital epidemiologists, employee safety officers, administrators on call, operations chief, and vice presidents for facilities and information technology. In sum and in our shop's shorthand for all of this, know where your flashlights are stowed.
11. Avoid the temptation—even if you're pressured by administrators or faculty—to broker crisis information to "friendly" journalists or withhold it from "unfriendly" journalists. Either strategy may feel good and get you somewhere for one news cycle, but eventually there will be a mess. One reporter may break a story by her

own enterprise, but once it's out there, the very essence of crisis communications means getting information out consistently, with an even hand, as soon as possible, to all media.

12. Never agree to bend the truth, even if it's for a worthy cause. Your job is to figure out the best, most accurate, most truthful way to tell what can legally and ethically be told about a crisis. If your institution asks you to do anything you're uncomfortable with, it's your job to (a) explain why they shouldn't and you won't and (b) advise your bosses that it's far better to say they won't comment than to lie.

SUMMING UP: CRISES COME AND GO

How we react to them is long remembered. Cleave to that mantra of telling bad news first, fast, and fully. Nurture the wits and courage to know the information you are getting is the truth and tell truth to power in the press or the boardroom. And keep the flashlights handy.

39

Government Agencies

COLLEEN HENRICHSEN

Colleen Henrichsen has been chief of the Office of Clinical Center Communications at the Clinical Center, National Institutes of Health, since 1990. She has worked at NIH for more than 23 years. She also worked in public relations and communications at the U.S. Chamber of Commerce, Brigham Young University, and the United States Congress. She graduated from Brigham Young University in 1973 in communications.

A medical resident was on duty at New York Hospital one night in 1979 when a 27-year-old security guard was admitted with a rare form of pneumonia. As inexperienced as the resident was, he knew that this very rare condition was usually diagnosed only in people with a history of cancer, organ transplantation, or other conditions involving immune system suppression. This otherwise healthy young man had none of those. Weeks later, when the resident presented this case at inner-city rounds, a number of hands shot up. These clinicians had seen similar cases. The resident's report of this New York City outbreak was one of three that formed the basis for the first published report of the disease we would come to know as AIDS.

Three years later, this physician, Dr. Henry Masur, arrived at the National Institutes of Health where he joined established NIH researchers already anxiously trying to understand this deadly new condition.

At key communications offices on the NIH campus, phones were ringing incessantly. Reporters all around the world wanted to know what NIH was doing about it. As public communicators, we were learning about the disease along with the researchers. Why did it seem to disproportionately affect gay men? Why were people with the disease dying from ordinary infections? We were learning the answers to these questions as they unfolded, translating what we learned into plain English, and getting the information out to the public. A prominent AIDS researcher came into the office of one of my colleagues, sat next to her, and made a simple drawing of how immune cells appeared to be affected by the new virus, explaining it to her at the same time scientists themselves were just beginning to understand it.

Dr. Masur is now chief of the Critical Care Medicine Department of the NIH Clinical Center. I covered his account of his first meeting with an AIDS patient for an NIH employee newsletter when he delivered the NIH Astute Clinician Lecture in 2002, which honors scientists who observe and investigate unusual clinical occurrences.

There are similar stories of diseases studied at all the NIH institutes. And communications professionals at other U.S. agencies—including the National Aeronautics and Space Administration (NASA), the Environmental Protection Agency (EPA), the National Science Foundation (NSF), and the Centers for Disease Control and Prevention (CDC)—tell similar tales of having to tell the story as the science or science policy unfolds. As a communications professional at a science-oriented government agency, you're not exactly part of the scientific process, but you are definitely along for the ride.

This is not a job for people who want to see their bylines in print. But it can be satisfying to have a role in disseminating scientific findings to those whose lives these findings will affect. Jobs for public communicators at science agencies may suit writers with a range of talents and temperaments, whether they are trained in science writing or are generalists with the curiosity, motivation, and aptitude to learn the science.

Where the Government Jobs Are

Many government agencies employ science writers. Most, like NIH, are congressionally mandated to make their research findings, policies, activities, or regulations accessible to the American taxpayer. In addition to NIH, NASA, EPA, NSF, and CDC, other science-oriented agencies include the Food and Drug Administration (FDA), the National Institute of Standards and Technology (NIST), the Smithsonian Institution, the U.S. Department of Agriculture (USDA), and the National Oceanic and Atmospheric Administration (NOAA).

Science-oriented congressional committees—and some members of Congress whose districts include high-tech industries—sometimes also employ science writers.

Media Relations

Government science agencies need a media-savvy cadre of communications professionals because politics inevitably plays a significant role in how the press covers government science. When clean air or clean water regulations are changed according to the philosophies of a specific administration, the EPA gets major media attention. NIH involvement in politically sensitive research, such as sexual behavior or embryonic stem cell cloning, can create a media frenzy. When a space shuttle explodes, killing the crew, or the Hubble Telescope malfunctions, it's a media event. When the FDA denies approval of a promising new treatment, it can have major financial consequences (think ImClone—while this company's treatment was eventually approved, its initial denial led to a financial scandal for Martha Stewart), and the media are all over it.

Because government scientists essentially work for the American public, government communicators have an obligation to engage proactively with the media and encourage scientists to talk to reporters. Scientists are often thrust into the public spotlight unintentionally and may be unprepared and unsure of how to respond to media inquiries. It is our responsibility to help them learn how best to interact with reporters and tell their stories of discovery in plain language.

What Makes News in Science

Science and biomedical advances published in top-tier peer-reviewed journals (for example, *Science, Nature Medicine, New England Journal of Medicine,* and *JAMA*) get the most newsprint and airtime. In medicine, the stories most likely to be covered are research advances on diseases that affect large numbers of people—for example, cancer, diabetes, and heart disease. Stories about celebrities also tend to create a sensation and draw interest to the diseases the celebrities have. Actor Michael J. Fox's announcement that he had Parkinson's disease generated many stories about that condition. Anything controversial or tragic or frightening creates news. The presence of mad cow disease in the United States drew hundreds of reporters.

Quotes from scientists, whether as part of a press release or as responses to inquiries, add credibility to stories. They also add clarity and interest and make the story easier to understand. Working in collaboration with scientists, government communicators determine whether a scientific finding warrants a press release or whether to simply make scientists available to reporters for comment.

When a significant scientific finding is likely to create potent media interest, organizing a press conference can provide a consistent, accurate message as well as save time and prevent confusion for everyone involved.

Handling Media Inquiries

Agencies differ on procedures for handling media inquires. Some require scientists and administrators to refer reporters to the institution's communications office. NIH encourages scientists to contact their institute's communications office. Because NIH is such a large organization, reporters often call several sources in several different NIH institutes about the same issue. By coordinating calls through the various communications offices, NIH can provide reporters with the best source and deliver a more consistent and accurate message.

NIH is also part of a larger agency, the Department of Health and Human Services (DHHS). The DHHS communications office is responsible for making sure that messages across its agencies (for example, NIH, FDA, and CDC) are coordinated and that the DHHS Secretary is informed of major media interests and activities.

Patient Interviews

When reporters are investigating a disease, they usually want interviews with patients who have the disease, to add human interest to the story. NIH operates a research hospital, the Clinical Center, where clinical research studies are conducted. The Clinical Center allows patients to be interviewed as long as both the patient (or patient's legal guardian) and physician agree to it. The public communications staff assures that permissions are obtained, that appropriate forms are signed, and that the patient understands the implications of speaking to the press.

Freedom of Information Act (FOIA)

The Federal Freedom of Information Act, passed by Congress in 1966 and amended periodically since, allows U.S. citizens and foreign national residents to request records from the executive branch of the federal government. Communications offices are frequently charged with coordinating FOIA requests because the act is often used by the news media to get government information. FOIA doesn't cover Congress, the federal courts, or state and local governments, but it does cover all Cabinet agencies, independent agencies, regulatory commissions, and government-owned corporations.

Records include all documents, papers, reports, and letters in the government's possession. The term "record" has also been ruled to cover films, photographs, sound recordings, and computer tapes. Records that can be released under FOIA include meeting minutes, e-mail messages, computer files, and document drafts.

There are nine exemptions to FOIA, which form the basis for withholding records. The two most often used exemptions are invasion of privacy (such as medical records) and commercial or financial information. Handling FOIA requests requires responding promptly and following the law to the letter.

Publications for the Public, Professional, and Employee Audiences

NIH public communications offices produce hundreds of publications and other communications materials for professional audiences and the general public. These materials take many forms, including books, pamphlets, fact sheets, generic video footage, videotapes, and exhibits. Most information materials are free and are not copyrighted. They are used in clinics, schools, and libraries and by individuals, making them a remarkable treasure for the public. Materials for the general public are usually focused on descriptions of, and treatments for, specific diseases. NIH materials cover every conceivable condition, including cancer, heart disease, diabetes, arthritis, mental illnesses, brain disorders, eye diseases, and drug abuse, to name only a few. Many publications are also education oriented and contain information on healthy behaviors or the science behind health—for example, what microbes are or how vaccines work.

Some of the larger NIH institutes produce national public education campaigns. These include the National Cancer Institute's long-running "Five a Day" campaign to encourage the consumption of fruits and vegetables to prevent cancer, and the Red Dress campaign begun in 2003 by the National Heart, Lung, and Blood Institute to raise awareness of heart disease in women.

Many agencies contract with clearinghouses and communications firms to both produce and distribute these materials. Science writers are hired not only as federal employees, but also as freelance contractors to produce government publications, health education campaigns, and other communications materials. Freelances are most valuable when they have a strong grasp of science as well as an understanding and appreciation of the agency culture.

The NIH Clinical Center produces a series of patient education publications, both in print and online, that describe the procedures patients undergo at the Clinical Center.

Some government agencies also produce employee communications. The Clinical Center produces a monthly hospital employee newsletter, for example, and the Office of the NIH Director produces an agencywide bimonthly employee newsletter. NIH also produces a monthly newsletter aimed specifically at keeping scientists informed about each other's work as well as about issues and policies bearing on the conduct of research.

Annual reports, research reports, and responses to congressional inquiries are often standard responsibilities of government communications offices. Policy offices often employ science writers to write more complicated policy analysis for Congress, remarks for congressional hearings, and statements for advisory councils.

Websites

One of the most dramatic changes in the field of communications has been the introduction of Web technology. Nowhere has the change in the way the public finds information been more dramatic than in the field of health information.

Many people now get their health information from the Web. A study by the Pew Internet and American Life Project reported in July 2003 that 80 percent of adults use the Internet to find health information. This makes writing for the Web an important avenue for science writers. It is especially significant for government science writers, because studies also show that people trust the health information they receive from the government.

A report by Consumer Webwatch published October 29, 2002, on building trust on the Web noted that their panel of experts ranked NIH's website as first in credibility in providing health information. People trust the information because they know that as a government agency, NIH has no commercial interest in promoting a particular treatment. The experts noted that the site references peer-reviewed journals and is often used as a source by other sites.

These studies make clear the importance of good science writing on government websites. In the early days of Web development, there was a general opinion among scientists and computer experts that with the Web, one had an unlimited resource for disseminating information. The Web wasn't subject to the space limitations of publications, or to the whims of journal editors, and it was free. As the technology has matured, it has become clear that, while the Web is a wonderful resource, limitations on time and resources make it imperfect. Because information on the Web can be updated immediately, expectations are that the information will be constantly current. But this requires having a system in place that prompts subject experts to review content regularly, and enough technical support to update the site daily.

Incidentally, Consumer Webwatch found that credibility wasn't based as much on surface issues, such as the quality of visual design, as on the quality of the sources. In fact, designs that were too flashy made the experts question whether the sites were more show than substance. Not surprisingly, the report found that poor grammar and typos made users question the site's authoritativeness.

Advantages/Challenges

One of the great advantages of being a science writer for a science-oriented government agency is working with scientists who are dedicated to (in the case of NIH) understanding and curing disease and relieving suffering in the world. One of the greatest challenges for government public information officers is that scientists sometimes don't appreciate the importance of making their findings available to the general public. They often are at odds with the news media. Reporters often are looking for the breaking story. But science doesn't usually break forth. It takes baby steps, over the course of years or generations. Some scientists fear that their findings will be contorted by the news media into something they're not, to make them more newsworthy. The professional communicator is the mediator between these two viewpoints. We can work with journalists to assure that the story is both significant and accurate. And good experience with the media can convince scientists that there is value in increasing the public's understanding of science.

And science does move forward, often in unexpected ways. Unlike other areas of government where policy and regulation can be the sole determining factors in career direction, in a science-oriented agency it is the novel and unpredictable events of science that can set the course of your career. In the 1970s no one could have predicted the impact AIDS would have on the nation or the careers of science writers. In the future, biodefense and the related issues of smallpox, anthrax, and ricin may take our careers in a new direction. Or new and emerging infections, such as SARS or monkeypox, may occupy our time. It is the unpredictability of science that makes careers in science agencies both challenging and satisfying.

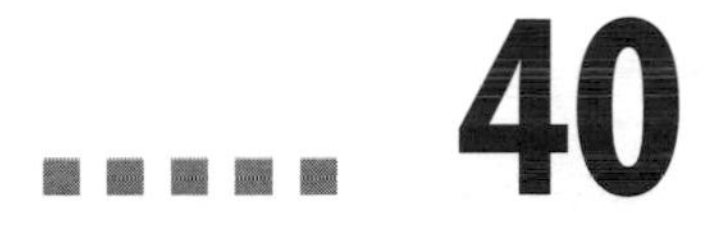

40

Nonprofits

FRANK BLANCHARD

Frank Blanchard joined the Whitaker Foundation in 1994 as director of communications, responsible for all outreach, including grants to the AAAS Science Journalism Awards and the National Press Foundation seminar series. He began his career in 1977 as an associate editor for the *Monroe Journal* in Alabama, moving to political writer for the *Montgomery Advertiser,* and then to night editor at the Associated Press in Atlanta. Frank left journalism in 1985 to become senior science writer for the University of Michigan news office. He joined the Howard Hughes Medical Institute as senior information officer in 1988. He holds a bachelor's degree in journalism from the University of Florida.

The door swung wide open and a dozen faces turned my way. The interview was for a science education story, but now it felt like a surprise party and I was the guest of honor. I had never been greeted for an interview by so many people. Here was a crowd around a long table. At the head was a woman with a big smile welcoming me to Xavier University of Louisiana in New Orleans.

My traveling companion, a consultant experienced in representing charities, private foundations, and other nonprofit organizations, took this in stride and spoke right up. He introduced us as representatives of the Howard Hughes Medical Institute in Bethesda, Maryland, which at the time was the nation's largest philanthropy, a $5.2 billion enterprise. The Institute had given this relatively small college, which led the nation in sending black students to medical school, a $1.8 million grant to support its undergraduate science program. Xavier, the only historically black Catholic university in the United States, was doing something extraordinary. We wanted to tell the story. As a science writer for the Institute, I had made arrangements to interview the faculty member who ran the program and a few participating students for anecdotes to enliven the story. I thought I had made this clear on the telephone weeks before the trip, but the program director had other ideas.

She had built an itinerary that could have stood as a first-class defense of the grant. Faculty members, support staff, anyone who could bolster the case for funding was in the room. After meeting with the group, I was to speak with the university president and others. later I would have a chance to talk with students. It was going to be a long day. It was also going to be a waste of time.

We were there to gather specific information for a news article. We were not there to monitor progress under the grant. We had no money to offer them, and we were in no position to cut their funding. Finally, as this began to sink in, the teachers headed back to their classrooms. Other staffers returned to their desks. the program director sat down with a pen and reworked the day's agenda.

For me, this was an early lesson in what it means to be a science writer in philanthropy. Any foundation representative who visits a grant recipient wears the mantle of the funding agency and carries the promise of continued support or the threat of a funding cut. You're no longer just a writer. You're someone with connections. You can always put in a good word with the right people.

Dual status as writer/foundation representative can interfere with getting things done. It can also have the reverse effect. It can open doors. While at the Whitaker Foundation, I had access to a key document in the early campaign to create what became the National Institute of Biomedical Imaging and Bioengineering at the National Institutes of Health. A 1995 consultant's report to Congress (www.becon.nih.gov/externalreport.htm) needed some work. A politically sensitive issue was how to ask for a new institute to be created without asking for a new institute to be created. We did some rewriting, which the consultants accepted and included in Recommendation 1: "NIH should establish a central focus for basic bioengineering research. This central focus should be at the highest level and should include resources for the collaborative support of extramural research."

Foundations and other nonprofits tend to be collaborative places in which ideas and creativity are encouraged. Nonprofits often attract people who are highly educated, passionate about a cause, have the public interest at heart, and love their work. Instead of making things, they make things happen. Nonprofits benefit from government tax breaks, a sort of public subsidy. The price for this financial freedom is accountability to the public for the use of these tax-free resources. In this sector, there are two major demands placed upon writers: to help accomplish the mission of change and to explain accomplishments to the public. These principles can sometimes bring a clear focus to what is written.

At a university, for example, a research story has several roles to fill, such as bringing recognition to the institution, the academic department, and the faculty member. A university news release might start this way:

> MADISON, Wis.—A new technology developed by a research group headed by Nimmi Ramanujam, assistant professor of biomedical engineering at the University of Wisconsin–Madison, will be a "third eye" during breast biopsies and can increase the chance for an accurate clinical diagnosis of breast cancer.

A foundation, on the other hand, might focus more intently on the accomplishment:

> ARLINGTON, Va.—A light-sensitive probe is being developed to help doctors spot breast cancer in some of the 70,000 American women each year whose malignancies fail to show up in needle biopsies.

Nonprofits nurture collaboration within and frequently collaborate with each other and with other organizations. These relationships sometimes force the writer into the role of negotiator. As the science writer for a funding agency, you may want to publicize the results of a grantee's research. The grantee's home institution wants to do the same, but with different motivations. As a result, the two accounts of the research may have different slants intended for different audiences. They may both end up with an overlapping readership, creating confusion. One strategy is to develop relationships with your counterparts at other organizations. You might suggest issuing a joint news release or coordinating dual releases, taking care to avoid unnecessary redundancies. For example, the funding agency may distribute to national news organizations, while the grantee concentrates on state and local coverage. If more than one funding agency and multiple collaborators are involved, the writing may suffer from too many fingers at the keyboard. You may have to rely on your negotiating skills to balance competing interests and scratch out a well-written piece.

In the nonprofit sector, the science writer routinely handles a vast array of assignments. You may write news releases, annual report articles, newsletter stories, brochures, policy briefings, white papers, talking points, meeting reports, research abstracts, website stories, occasional papers, or pitch letters to journalists. It may be up to you to develop a survey instrument and then report the survey results. You may be assigned to write speeches for the board chairman. You may find yourself on the front line of the grant-making process, conceiving and writing grant announcements and requests for proposals. Your nonprofit may instead have to raise money, and you would write grant applications. You may lead workshops with constituency members, encouraging them to volunteer as sources for your stories. You may do all of these and more.

At the same time, there are jobs that have more of a journalistic flavor. Joan Arehart-Treichel left *Science News* for a position with *Psychiatric News*, published by the American Psychiatric Association, a nonprofit membership organization. She says she's doing pretty much what she did before. "Most of the time I report scientific developments straight from the journals and talking to scientists. It brings me great joy not having to get into the politics. . . . I follow a lot of journals online. I like going fishing to see what's new and what's important." She originates most of her stories, using conventional news judgment, and occasionally gets to take on topical pieces, such as depression in college students and the unique challenges that face psychiatrists abroad: "I was in the Arctic last summer doing a piece on practicing psychiatry at the top of the world."

Joan's parent organization does have political issues. Some occasionally trickle down to the magazine. Psychiatrists and psychologists are currently at odds over whether psychologists should have the authority to write prescriptions. This means that a research study by a psychologist may not find its way into *Psychiatric News*. But Joan says it's more likely that a story will get spiked because it relies too heavily on investigator speculation and too little on published results.

Every organization has its reasons for publishing. Nonprofits usually focus on what they are trying to accomplish and on publicly accounting for their progress. So in addition to having a good story to tell or publishing something purely in the public interest, there is a concurrent set of motivations at work. This presents the science writer with some interesting challenges.

William Stolzenburg, science editor of the Nature Conservancy's flagship publication, *Nature Conservancy*, must consider the organization's science-based, nonconfrontational mission to preserve the world's ecological diversity. A newspaper might focus on the destruction angle. Stolzenburg is compelled to do otherwise. "For example, a feature on prairie dog conservation avoids the most obvious tack of detailing the cruelty and persecution that has brought the animal to the brink of ecological extinction," he says. "Instead of getting inside the heads of those whose concept of good sport is blowing prairie dogs to smithereens, the story explores the latest ecological studies that show the prairie dog as keystone species of the North American grasslands, a dawning perception that has the beleaguered rodent winning overdue respect."

Nonprofit research institutions often incorporate public education into their communications priorities. The challenge for nonprofit science writers is to convince their organizations that communication precedes education. Rick Borchelt, director of communications Berman Bioethics Institute, Johns Hopkins University, recalls a conversation with the creators of a high-energy physics exhibit that traveled through Europe. In evaluating the exhibit's effectiveness, the creators learned that the written material was too complex for visitors to understand. "One conclusion you could draw from that is the need to write to the level of your intended audience," Borchelt says. "But the conclusion the team drew was that physics education needed to be better so that audiences could understand their exhibit.

"This is a commonly held—if uncommonly articulated—philosophy among nonprofits and advocacy organizations," he says. "And it presents a real challenge to science communicators who know that tailoring a message to existing audience skills and knowledge levels is key to effective communication. It's a lesson we would do well to learn from the commercial advertising sector."

Issue-oriented nonprofits often have a national perspective on advances in science and engineering. Research institutes can place an otherwise isolated advance into a national context. Funding agencies can thread together research from numerous institutions under a single topic, adding credibility with sources and experts from a wide range of institutions nationwide. This gives the nonprofit science writer more flexibility than the corporate or university writer, who may see other institutions as competitors.

For journalists, there is often a strong drive to get the story, to follow a hot lead and not let go. The writer at a nonprofit must be prepared to give up the good story. The American Chemical Society (ACS) prints about 21,000 articles a year in 34 publications. It's impossible for a single writing staff to give this volume of material the news coverage it deserves, even by skimming the best stories from the top of the heap. The ACS has managed this mountain of material by handing off stories to writers at academic, corporate, and government offices. These outside writers get to work on exciting research stories, which they issue as news releases from their own institutions.

"You give the public information officers as much information as possible," says Denise Graveline, former director of communications for the ACS. "They extend your reach. It's one of the best-kept secrets in the universe. Everyone gets a chance to contribute. It's fabulous."

Opportunities for employment in the nonprofit sector are on the rise. The Independent Sector, a Washington, D.C.–based coalition of leading nonprofits and foundations, reports that the number of Americans working for nonprofits has doubled in the past 25 years, climbing to 12.5 million, or 9.5 percent of total U.S. employment. The Foundation Center in New York counts more than 50,000 foundations in the United States. Among the largest, 800 are involved in medical research, 200 in science, and 200 in engineering and technology. Hundreds of other foundations support health care, the environment, and other areas of science and technology. In addition, there are countless nonprofit research institutions, museums, science centers, science education organizations, funding agencies, science policy institutes, and other nonprofits that hire science writers on staff or engage their freelance services.

Many nonprofit employees contend that it is easy to achieve high levels of responsibility, deal directly with upper management and top leadership, and achieve success early in a career. According to the University of Delaware, nonprofits offer more opportunity for creativity, diversity in job assignments, and flexibility in schedules. New graduates can gain broad experience in a short time. The downside may be lower salaries, lax organizational discipline, less long-range planning, and financial instability. Even so, the rewards can satisfy. Danette St. Onge told *Philanthropy News Digest* that she had passed up several high-paying tech companies a few years back to manage the nonprofit Exploratorium in San Francisco. In addition to finding the new work more meaningful and filled with purpose, she "delighted in the fact that four out of five of the dot-com companies . . . were bankrupt within a few months."

41

Museums

MARY MILLER

After graduating from UC–Santa Cruz with a degree in biology and a master's certificate in science writing, Mary Miller assumed she would get a job at a small newspaper, the recommended route for aspiring science journalists. Instead, she heard of an opening for a science writer at the Exploratorium, the famed interactive science museum. An adviser told her it probably wouldn't hurt her career to "play at the Exploratorium" for a year or two. In the dozen-plus years since happily landing that job, she says she's never been bored and never regretted her choice of career. During this time, Mary has also freelanced for such magazines as *Natural History, New Scientist, Smithsonian, Popular Science, California Wild,* and *The Sciences,* and on numerous websites. She co-authored the book Watching Weather (1998) and contributed to two Discovery Insight Guides, *Dinosaur Digs* (1999) and *Scuba Diving in North America and the Caribbean* (2000). She is past president of the Northern California Science Writers Association.

Science writers at a museum, zoo, or aquarium are in a powerful position. We provide the first line of information that visitors receive about the place. The reading public comes eager to be inspired or entertained and maybe learn something about science and nature in the process.

One of the most important jobs for a museum science writer is producing the text that accompanies exhibits. Exhibit writing was once the province of scientists or specialist curators, who felt no guilt about putting up dense technical prose for the visitor to either plod through or ignore. As long as the label didn't misidentify a dinosaur or a physical law of nature, all was well.

Thankfully, the last 20 years have seen an evolution in museum exhibit writing. Curators and museum directors began to take pity on the visitor and started hiring professional writers to make the museum experience less mystifying. Museum developers have become aware they are not talking to themselves, but to an audience that might need some help understanding the physics exhibit, stuffed animal, or strange deep-sea jellyfish swimming in front of their eyes. It can be a challenge, especially at a museum like the Exploratorium, where successful interactive exhibits must be both operated and understood by the visitor.

Few writers have so many functions to serve in so few words. A title and a tag line might call on the kinds of skills an advertising copywriter has, pulling people in before they know what they're going to be doing. Then a set of instructions helps a visitor build, experience, or do something that may or may not "work." After that, you get to be a narrative science writer, explaining what just happened and why, translating, for instance, from the point of view of a biologist, physicist, or exhibit builder. Next, you might turn into a social commentator or a science historian, connecting the experience to the real world or pointing out the exhibit's historical significance. All in no more than 100 words, shorter than this paragraph. It's a tough job, but it can be rewarding when all the pieces come together.

Increasingly, this job is no longer the sole responsibility of the writer or editor. In today's evolved museum, exhibit text is often produced during a drawn-out exhibition development process (an exhibition is a group of themed exhibits with explanatory text and graphics). The writer is a member of a design team that researches a topic, argues about communication goals, conceptualizes individual exhibits and their accompanying labels, evaluates their effectiveness with visitors, and repeats the process as many times as needed before the final exhibition goes on the museum floor. Each step, including the text writing, goes through this incremental process, with everyone on the team weighing in on every widget and knob and every noun, verb, and comma. It can be like writing a novel by committee.

And like writing a novel, the exhibition development process can consume years of your life, typically three to five years from start to finish. In the end, all the writer has to show for the effort is a few thousand words that have been picked over and changed so many times that you might not recognize your own writing. When successful, the individual exhibits can teach and delight museum visitors for years. But, like a newspaper or magazine journalist who feels wounded by the editing process, sometimes we exhibit writers fret that our work has been dumbed down or that our individual voice has been drained from the text and replaced with an institutional voice that lacks sparkle or personality. The process itself can also be frustrating for lone-wolf writers, who tend to be product rather than process oriented. But the process is important, if tortuous, because exhibit labels have more permanence than other forms of writing. Our labored-over words are often etched, laminated, or engraved and can remain in place for decades, a very sobering thought.

All good writers care about their readers, but museum writers have a special relationship to their audience. How many writers can regularly watch folks reading their words? We have only to stroll out on the exhibit floor to see our readers. As a result, we tend to have great empathy for our audience, who must, after all, read our work standing up. Exhibit writers pack maximum meaning into minimal words; it's the haiku of science writing. We must give visitors just enough explanation to both operate the exhibit and understand a little about the science behind it. The exhibit label usually consists of short declarative sentences: *Turn that crank. Push the red lever. Notice what happens.* It's hard to write so sparingly, but it's critical for an exhibit label. Faced with too much text, visitors will often walk away without touching the exhibit. Too little explanation, and they're confused about how it works or the science behind it. To add another layer of difficulty, we're designing exhibits and writing for an audience that ranges from 6 to 86. The target audience for science museums, zoos, or aquariums is broad, multigenerational, and demanding.

Museum as a Teaching Tool

Fortunately for science writers who like to write, there are plenty of other words that need to be produced in a museum. Every piece of written text, from the mundane signs directing visitors to the ticket booth or bathroom, to exhibition catalogues, newsletters, press releases, posters, books, magazine articles, ad copy, classroom guides, websites, grant applications, and reports must be authored. I write about weather, astronomy, particle physics, biology, global climate change, human evolution, wine, and music. In the last few years, my museum job has taken a multimedia, globetrotting turn. I've produced and hosted expeditions and webcasts that included such adventures as diving under the ice in Antarctica, donning a bunny suit to host a live webcast inside the gigantic clean room at Goddard Space Flight Center, and interviewing the giants of DNA discoveries at Cold Spring Harbor Laboratory.

One of my colleagues who worked at Sea World in San Diego says that the most widely read copy she ever wrote was 50 words about sea otters printed on the paper cups used in the cafeteria. After nearly three decades and millions of soda cups, her educational prose about an adorable marine mammal lives on.

From soda cups in the cafe to exhibits on the floor, museums are considered important locations for informal education. Museums such as the Exploratorium also serve the needs of formal education, in the form of school field trips and teacher training and support. Nowadays, museum field trips are taken seriously as an educational opportunity rather than simply a chance for kids to blow off steam. To help teachers prepare for a field trip, Exploratorium scientists, educators, and staff writers create activities and background materials to be used by school groups both before and after their visits. The materials are written and tested in the trenches with teachers and kids. Many of our favorite activities are weeded out in the process. One activity, in which a student uses her forearm and hand to estimate the height of objects, went through many written incarnations before we realized that it simply didn't work. Although it could easily be demonstrated in front of a classroom with an experienced guide, we couldn't make this activity work on the printed page, even with detailed illustrations.

The Virtual Museum

Sometimes an animation or video clip shows what can't easily be told. To expand the visitor experience, the Exploratorium and other museums are experimenting with video, handheld devices, interactive kiosks, and other electronic media on the floor. These create more opportunities for writers in the form of video and animation treatments and scripts and written background for visitors to go deeper into the exhibit experience.

As museums and other institutions go beyond the walls of their buildings and expand into the online world, the Web becomes another arena for the museum science writer. The Exploratorium was one of the first public institutions on the World Wide Web, establishing our presence there in 1994. From the beginning, we used our website not simply as a marketing tool but as an opportunity for creating original content that ranged from online exhibits to Web pages. We explored such diverse content as the science of sports, the perception of wine, and the phenomenon of space weather. As we do for our museum visitors, we cover the science of everyday life for online visitors curious about the world they live in.

The Internet also allows us to do what can't be done with exhibits on the museum floor: cover current science. The Web is a dynamic medium that can keep up with changing scientific information and understanding. For a website about global warming, called the Global Climate Research Explorer (www.exploratorium.edu/climate), we created a front page that incorporates real-time satellite images and updated temperature graphs and charts. We've pulled content from different research organizations and universities involved in studying the changing world climate, and provided the means for people to interpret and understand the data that scientists themselves gather and evaluate.

In this capacity, we are acting as mediators between the formal world of science and the general public. Since global warming is a hotly debated topic, we felt it was important to show real data and evidence, not just interpretation, even if it meant displaying complex maps and graphs.

From Writer to Producer and On-Camera Talent

One of the most exciting developments in my career at the Exploratorium was a project called Origins (www.exploratorium.edu/origins). This NSF-funded endeavor allowed us to travel to six scientific laboratories or observatories, from the particle accelerator at CERN to the Antarctica research stations at McMurdo and the South Pole. We created virtual field trips to these locations, giving our audience a behind-the-scenes glimpse of the people and process of science. To bring our visitors along on the journey, we used traditional storytelling and the tools of multimedia: text, photographs, video, audio, interactive animation, and live webcasts.

As a producer and one of the project leaders, I expanded beyond my writing niche and learned to tell stories in different ways. I learned to shoot and edit video, crop digital photos, and paste together Web pages. I carried video and still cameras, microphones, and tape recorders, along with my traditional notebook and pen. I picked up these new skills on the job and under deadline pressure, a scary but extremely effective way to learn.

One of the most nerve-wracking aspects of my job, at least at first, was producing and acting as on-camera host and interviewer for the webcasts. Webcasts are programs, often in front of an audience, that are streamed live on the Internet. Because it's live, anything can happen, as I learned during our mummy webcast. We had arranged for a CT scan to be performed on an Egyptian mummy that we had borrowed from a museum. This mummy had never been medically examined and, from the writing on her coffin, was believed to be Princess Hatason. But, as we transmitted live images of the body beneath the wrappings, the radiologist pointed out that our mummy had a penis and therefore must not be a princess, but a prince—much to the delight of the kids in our studio audience.

For the Antarctica expedition, we did at least one webcast a day for six weeks, so I soon learned to relax in front of the camera despite the potential for technical difficulties or fear of getting tongue-tied. As long as there was a scientist guest for the webcast, I could retreat into my journalist mode and just ask questions. I would prepare for these webcasts by doing some reading and research about the scientist's work, spend some time talking to him or her, do a quick write-up for the website, and prepare a list of questions. Once we went live, I depended on my interviewing skills to get me through the webcast. We included digital photos or video clips when they were available, giving the audience an interesting image besides talking heads.

We also traveled to research locations away from the stations, took pictures and video, and wrote articles for the website. For me, Antarctica was the adventure of a lifetime, made all the better because I was paid for it. In addition to diving under the ice, I slept in a tent next to a groaning Dry Valleys glacier, shot video in ice tunnels 50 feet below the South Pole, spent a day on a Coast Guard ice breaker trailed by orcas, and hung out with Emperor penguins under the midnight sun. With experiences so rich, the writing came easily.

I hope it's obvious by now that working for a museum can be an incredible learning experience for an enterprising writer. My job at the Exploratorium has given me the chance to meet and interview some of the superstars of science, from James Watson to Brian Greene to E. O. Wilson. I've shaped my job by always looking for new opportunities, such as learning to shoot video, take good photographs, and seek out new markets. Two things that have made a real difference were getting involved in grant writing and keeping up with current science by attending conferences such as the American Association for the Advancement of Science and the American Geophysical Union.

New projects always need funding, and so they're often decided in the grant-writing phase. When we were writing the Origins grant to NSF, I convinced my boss that Antarctica was a viable location for us even though no other museum group had ever gone there. I wrote the application and lined up support from NSF's Office of Polar Programs, Artist and Writers Program. Thankfully OPP's Guy Guthridge shared our vision, and I became the project leader for the Exploratorium team.

A science writer going to conferences seems like a no-brainer, but I rarely see my museum colleagues there. I've lined up valuable contacts and heard about big projects coming down the pike, all of which I've used to advantage in my Exploratorium job (not to mention lining up an occasional freelance gig as well).

With the increasing realization that museums play a vital role in the public understanding of science and technology, funding agencies and visitors alike continue to support our work. The good news is that museums are no longer considered the backwaters of science writing, but a viable and expanding career choice for our field.

42

Corporate Public Relations

MARION E. GLICK

Marion E. Glick is a senior vice president of health care media relations at Porter Novelli, one of the world's largest public relations firms. She began her career at the Johns Hopkins Medical Institutions, where she was the first HIV/AIDS spokesperson. She then became chief of the Information Projects Section in the Office of Communications at the National Institute of Allergy and Infectious Diseases, part of the National Institutes of Health. She was director of communications for the Rockefeller University before moving to Porter Novelli. Marion holds a master's degree in journalism/science communication from the University of Maryland College of Journalism and a bachelor's degree in biology from Muhlenberg College, where she is a member of the board of trustees.

Public relations is not just sending out a press release or invitations to an event. It is the profession of managing communications between an organization and its audiences. As a public relations professional, you develop and execute communications programs that consider and support such corporate goals as reputation, the selling of products or services, recruitment of employees, or encouragement of investments. You can do this as an in-house professional at the company or as a client service if you work in an agency.

If you want to apply your science journalism skills to corporate public relations, they will be highly prized by pharmaceutical, biotechnology, medical device, technology, and related companies. You not only comprehend the facts about environmental, physical, or life sciences, you can make them understandable to others. You can accurately and efficiently translate the function and value of a product or service to audiences as varied as customers, stockholders, regulators, and journalists, all of whom have different levels of scientific understanding.

But being savvy about the scientific process and journalism is not enough. You also have to understand the business. Yes, it is about the money or, rather, commercial decision-making. To do your job well, you must know how the company makes money, who runs the show, who are the customers, how the business will grow, how it is regulated, and who are the existing or potential partners and competitors. And you should know these aspects as well as you know the company's research and development pipeline, patents, or marketed products or services.

As someone who made the transition from managing public relations about medical research for academic and governmental organizations to that of pharmaceuticals and biotechs, I can say that mastering "the business stuff" is possible. Many excellent resources are available, but start by skimming business magazines, checking out Hoover's Online (www.hoovers.com), and reading the annual reports of your company or clients.

To manage corporate public relations, you need a program, which is the blueprint that captures the vision *and* the means to obtain it. Programs are very structured and have goals, objectives, strategies, and measurable tactics to achieve them. This structure allows planning of staffing, budgets, and timing. The program also must determine and measure expected outcomes, that is, "what success looks like," because public relations must be accountable.

For example, the corporate goal might be to sell a new, first-in-class cholesterol-lowering drug. A public relations objective would be to increase awareness of the drug's significant efficacy and exclusive mechanism of action. A strategy would be publicizing drug trials presented at the American Heart Association meeting. Basic tactics then would be to write a news release about the studies, develop a list of media for the release's distribution, pitch the news to reporters, arrange interviews, and monitor media coverage. The monitoring provides a measure of both the quantity and quality of your efforts.

Many journalists move into public relations as freelances or staffers who write or produce corporate press releases, media alerts, video news releases, speeches, question and answer documents, annual reports, op-eds, or articles for in-house outlets.

In the corporate world, your colleagues may not have your understanding of journalism. The product manager may not know that headlines are written by editors, not reporters. The technology transfer director may be clueless about journal embargoes. The marketing director could be an ace about ad rates for the *New York Times* but not know the difference between the Associated Press wire and PR Newswire. Public relations professionals must master this media information.

You also know how journalists judge news, and as a public relations manager, you must be able to explain the difference between real news and hype or fluff to help corporate directors understand how journalists' perceptions shape coverage, or lack thereof, of company announcements. When discussing a media relations strategy, you may find yourself mentioning personal information about journalists to explain how you will garner their interest, known as the art of "pitching." But a word of caution: One of the most overrated assets in a public relations professional is knowing journalists personally. I'm not talking about knowing a journalist's beat or deadlines—you must know these to target and pitch appropriately. Rather, some clients presume that my being acquainted with a TV network correspondent means instant news coverage. This assumption reveals their naiveté about journalism. The best contacts in the world will not get a dud item onto the front page.

Telling your managers or clients this and other information they may not want to hear takes tact and diplomacy, two essential skills for public relations. Others are integrity, flexibility, patience, composure under pressure, clear thinking, and organizational skills. As you get to know your clients, you will know how informal to be, but the basic rule is to always be more formal than the client.

In the end, what matters for making news pretty much is, as Joe Friday said: "Just the facts." In this era of consolidation, corporate news often involves mergers, acquisitions, and licensing. If the company is public, then its financing, earnings and senior personnel changes are news. Each step in the drug approval process is news for a smaller biotech, but not necessarily for a large pharmaceutical company; but the approval of a first-in-class drug or device is news regardless of company size.

I've focused on media relations because it's what I do, and I think this aspect of public relations is closest to science journalism. But public relations also involves issues management, often called crisis communications (as discussed in chapter 38), as well as other proactive, "grab attention" strategies and tactics. For this you need to know your audiences: Who are they? What media do they consume? To what professional organizations do they belong? How do they spend their leisure time? This information is important for a complete public relations program.

For example, let's look at a few program elements for launching a new treatment for people with end-stage renal disease, when the kidneys fail and the body retains fluid and harmful wastes. First, because diet is critical for such patients, you need to provide the drug's efficacy, safety, and dosing information not only to doctors and nurses but also to dieticians. You can do this via trade media coverage and "meet the expert" events at professional meetings (except certified continuing medical education that usually is managed by firms specializing in such courses). Second, because men are well known not to be proactive about their health care, we often try to reach them via the women in their lives. So target women's magazines with easy-to-understand information about the drug. Finally, demographically, minorities make up a significant proportion of end-stage renal disease patients. So educational activities for these communities are very important, such as collaborating with the American Association of Kidney Patients to reach Hispanic patients with Spanish-language materials on the disease and treatment options.

In corporate public relations, you can't be the lone ranger you might have been as a science journalist. You may be the only science expert in the public relations office, but you are part of a team. So who are the others?

If your company is public, you will work with investor relations specialists, who also are communications professionals. I believe that a strong partnership between public relations and investor relations teams is critical for a corporation to succeed. Although the primary investor relations audiences are bankers, analysts, and shareholders, public relations and investor relations often overlap in strategies and tactics to serve the corporate business goals.

You will also work with lawyers, who must balance the risks of communications with the benefits. They evaluate what must be disclosed while regarding what remains private, considering Food and Drug Administration and Securities and Exchange Commission regulations as well as intellectual property, for example. Hence, they review almost every public communication. The lawyers often are the ultimate editors, and they don't know AP style. I will be blunt. You cannot be wedded to your words, active voice, or inverted-pyramid style in the corporate world.

My best advice: Learn the corporate process, be involved in message development so you understand the issues before you write, offer sound and objective reasons for your text, and then give up ownership. It's not you versus them. It's us.

Other potential public relations partners will depend on your strategies and tactics. You may work with the government relations staff to plan a way to lobby for medical coverage for a new drug. You may collaborate with a medical liaison to identify which of the many clinical trial investigators would make a good spokesperson to discuss an investigational drug. You may work with third parties devoted to the disease your company's drug addresses.

For example, one of my clients provided an unrestricted education grant to a national patient organization devoted to mental health, which used the money to host a meeting for regional patient groups to discuss the need for new treatment guidelines for depression. Another client worked closely with several patient groups to communicate to their memberships the opportunity to enroll in a clinical trial to treat spinal cord injuries. Many clients routinely include patient groups as well as professional medical associations in their plans to share facts about clinical trial outcomes and drug approval status, because patients routinely turn to patient organizations first for information. While some of my clients have great relationships with such groups, others use my firm to act as a matchmaker for introductions and networking.

As a member of Porter Novelli, a global public relations agency, I can tell you that my life is busier than in my previous jobs. I work for several clients, and that takes great flexibility. My work can range from arranging dinner for journalists to meet a corporate spokesperson to helping plan all of the communications activities for one pharmaceutical company's entire oncology franchise, both drugs on the market and those in development.

I work when my clients need me, which can include evenings, weekends, and holidays, such as when the American Urological Association meets over Memorial Day. I participate in global conference calls for Europe- and Asia-based clients during their work hours. When the FDA approves my client's drug at 4:30 p.m., my colleagues and I can work late into the night to get the news out both to media and third parties, so they have accurate information when patients call the next day.

I travel at least once every two to three weeks to visit existing clients, pitch programs to potential clients, or attend medical or financial meetings. Some of my travel is a day trip to New Jersey, but it can also mean a week at a convention center. Most of my travel is within the United States, but if the work has a global aspect, so might my travel.

I still write press releases, backgrounders, and Q&As, but I am more likely to hire a freelance or edit the work of a colleague. I spend more time providing counsel, developing programs, going to FDA hearings, setting up media interviews at medical meetings, and working with the public relations staff of major journals like *Science* to coordinate my clients' publicity efforts that abide by the journal's embargo policies. I also help clients prepare for interviews and speeches. I like the variety because it means my career is never boring.

So after this snapshot of public relations life, why pursue this career?

- Because you will be constantly challenged to learn new things. My life apart from the lab bench is not one apart from science.
- Because your familiarity with the science writing community makes the job efficient. I decide to whom to pitch a story based on my experience, not opening a directory and calling everyone.
- Because you can be part of the team. At a large agency, I have colleagues with whom to brainstorm ideas, collaborate on projects, and

share the workload as well as the success. It's a safety net for answering questions like "What am I not thinking about?" and "How can I do this better?"

- Because you can ultimately help people. Being part of a company that can positively affect the public's health is a good thing. My clients help create diagnostics and treatments that my grandparents did not have, and I can't wait to see what proteomics brings!

And because it can be fun.

Additional Resources

To learn more, check out basic public relations textbooks, like those by Larissa A. Grunig, James Grunig, Scott M. Cutlip, Doug Newsom, or Fraser P. Seitel. And visit the following websites:

The Public Relations Society of America is the largest professional organization for public relations practitioners, including corporate, government, and nonprofit organizations: www.prsa.org

The Council of Public Relations Firms advocates public relations as a strategic business tool, promotes careers, and assists in setting professional standards: www.prfirms.org

O'Dwyers PR News Daily covers the industry and has a job center: www.odwyerpr.com

The Pharmaceutical Research and Manufacturers of America (PhRMA) represents research-based U.S. pharmaceutical and biotechnology companies: www.phrma.org

BIO represents more than 1,000 biotechnology companies, academic institutions, regional biotechnology centers, and related organizations in 34 nations; BIO members are involved in the research and development of health care, agricultural, industrial, and environmental products: www.bio.org

The National Investor Relations Institute is a professional association of corporate officers and investor relations consultants: www.niri.org

Epilogue

JAMES GLEICK

James Gleick, a sometime reporter, editor, and columnist for the *New York Times*, is the author, most recently, of *Isaac Newton*, a 2004 Pulitzer Prize finalist. His other works include *Chaos: Making a New Science* (1987), *Genius: The Life and Science of Richard Feynman* (1992), and several books on the interplay of technology and culture, which is also the theme of his website, www.around.com. He was born in New York City in 1954 and educated at Harvard, and now he lives in the Hudson Valley with his wife and two dogs. He is working on another book, about information and Information Theory, and continues to commit journalism from time to time.

In a magazine interview once, Marshall McLuhan—the great sage of mass communication, author of *Understanding Media* and *The Gutenberg Galaxy*—was going on about astrology, clairvoyance, and the occult. The interviewer asked whether he wasn't getting just a bit *mystical.*

"Yes—" McLuhan replied, "as mystical as the most advanced theories of modern nuclear physics. Mysticism is just tomorrow's science dreamed today."

A lovely aphorism—and utterly wrong. As roads to knowledge, mysticism and nuclear physics could not be more different. We citizens of the modern world desperately need to keep track of the difference, and this means remembering that science has a unique place in our culture. It has a special claim on the truth. What a nuclear physicist discovers may be wrong. It will be questioned; it must be provisional. And yet it deserves a kind of authority that should be denied to mystics.

We know this. Even people most hostile to science believe, in their hearts, that physicists are on to something. We've seen those bombs; we board jet planes; we carry cell phones. We know that such devices are powered by something more reliable than magic.

But McLuhan was on to something too. What he meant was that nuclear physics and the occult share certain difficulties. They are hard for laypeople to grasp. They employ esoteric language and obscure techniques. They produce results that seem miraculous. But here the similarity ends. Because the effects of science, no matter how startling, no matter how wonderful, are not miracles. This is why science writing is so hard—and why it matters so much.

■ ■ ■

Consider the following facts, learned by humans during our brief sojourn in this world, without the help of mystics or theologians:

1. Earth is one of several planets orbiting the Sun.
2. All species, including our own, evolved from earlier forms of life.
3. The global climate is warming, at least partly because of gases emitted by human activity.
4. Condoms help prevent the sexual transmission of disease.

All of these have been debated at one time or another, but they are true nonetheless. They have been through the fire: rigorous testing by the institutions and procedures of science. No serious person denies them—except, that is, persons with an ideological ax to grind.

How unsettling, then, that in the opening years of the twenty-first century, facts 2, 3, and 4 all came under attack by the government of the nation most renowned for scientific achievement. In the fourth century after Newton, not only does irrationalism wax and spread, it does so, at times, with powerful help from the press and, at times, under official auspices.

We were supposed to have made better progress by now. As long ago as 1922, President Woodrow Wilson said: "Of course, like every other man of intelligence and education, I do believe in organic evolution. It surprises me that at this late date such questions should be raised." Almost a century later, George W. Bush, during his first election campaign, said he neither believed in evolution nor disbelieved in it. His administration then lined up against it.

Here's how Bush's cabinet fought against evolution: In 2003, a biology professor at Texas Tech University declined to write letters of recommendation for students who did not accept the theory of evolution. The professor, who happened to be a devout Christian, said he felt he had a responsibility not to encourage the professional careers of students who chose to deny the basic methods and tenets of their discipline. (Anyone has a right to believe that the Sun revolves around Earth, but it doesn't bode well for a career in astrophysics.) The Bush administration wouldn't stand for this. Prompted by right-wing evangelical activists, the Justice Department undertook a formal antidiscrimination investigation against the professor. He was forced to back down.

This, of course, was merely one episode in a great drama taking place across the country: the conflict between religious pressure groups and some state governments on the one hand and scientists on the other. Oklahoma, Alabama, Ohio, Georgia, Texas, and Missouri were among the states whose legislators inserted themselves into the teaching of science for the purpose of advancing the mumbo-jumbo known as "creationism" or "intelligent design theory."

Mumbo-jumbo, meanwhile, is not one of the terms journalists are taught to use in their objective coverage of public officials. What's a good science reporter to do? What's the proper rhetorical style in covering this modern incarnation of the trials of Galileo?

Part of the answer is to report on the scientific process itself: the testing and questioning, the failures as much as the successes, the clashes of ideas and personalities. Another part is to make connections among different specialties—to remember that science is not merely a collection of facts, caged like the animals in a grand zoo, but an intricate, interconnected edifice. The fact of the heliocentric solar system cannot be plucked out; since Newton, it has been tightly bound to a vast body of understanding, a "system of the universe," gravity and the laws of motion, the oceanic tides and the flight of projectiles and all the rest. In the same way, evolution is not just a story about the past; it now informs modern medicine, genetics, and epidemiology, not to mention more distant realms, like computational ecology.

■ ■ ■

Oddly enough, we often use the same expression, *believe in*, for both science and theology. We say that we do or do not *believe in* evolution (or global warming or the Big Bang) just as we do or do not believe in God. It's an act of faith either way. Most people have little choice but to trust in the authority of scientists almost the same way they used to trust in the authority of divines. Lay readers are with Marshall McLuhan: Not fully understanding nuclear physics, how can they really judge its veracity?

In contrast to evolution, global warming is not a religious matter. It is an economic one. The most prominent spokesmen for the view that global warming does not exist are those financed by industries with a financial stake in the matter.

Climate change is about numbers and probabilities in a realm of noisy and chaotic statistics. When scientists estimate (as in a late 2003 study in *Science* magazine) that by the year 2100, global temperatures will have risen by 1.7 to 4.9 degrees Celsius above 1900 levels, it's no more than a guess about the future; they state their chance of accuracy, even within that relatively large 3.2-degree range, at only 90 percent. It's the best guess available, based on the broadest possible collection of data and the most sophisticated computer modeling, but it's by definition uncertain.

And yet, when the same scientists conclude, "There is no doubt that the composition of the atmosphere is changing because of human activities, and today greenhouse gases are the largest human influence on global climate" and that "significant further climate change is guaranteed," they state facts—facts for all their acknowledged uncertainty. That is science, after all: truth and doubt together. The challenge is how to convey this in the language of journalism.

■ ■ ■

Faith and credulity, skepticism and trust—these are core issues for all news reporters. For science writers, there is an extra layer of difficulty. It's hard to communicate well about science in a culture that continually outpaces its own brainpower with its technology. The connectedness of our world—our ability to communicate globally with unprecedented speed and intensity—has some unexpected consequences. One of the consequences may be a new form of mass hysteria. Example: the recent craziness that was known as Y2K—responsible authorities issuing warnings that built to a crescendo in 1999 of millennial power failures, bank panics, food shortages, and planes falling from the sky. It's hard to remember now what a fever pitch developed over this mostly illusory crisis. People all over America stocked up on bottled water and ammunition. The State of Ohio moved its emergency government operations into an underground bunker eight miles north of Columbus. Journalism, I believe, let us down. It fed on itself, channel upon channel of recursive self-reference. There was a failure of skepticism.

For better or worse, science writers have to serve as gatekeepers. They may need to be advocates, not for scientists, but for science, as a set of principles. This can be an uncomfortable role in a profession meant to value neutrality and balance, dispassion and an open mind. At some point, neutrality has to give way to common sense. Public opinion surveys continue to find widespread *belief in* psychic healing, extrasensory perception, ghosts, and clairvoyance. Along with creationism, these are absurdities, of course. Or if you don't like that word, try poppycock, self-delusion, moonshine, and twaddle. Yet they all have articulate and well-dressed spokesmen. If science writers don't separate the truth from all the bunkum, who will?

Index